KEW OBSERVATORY AND THE EVOLUTION OF VICTORIAN SCIENCE
1840–1910

SCIENCE AND CULTURE IN THE NINETEENTH CENTURY
BERNARD LIGHTMAN, EDITOR

Kew Observatory

& The Evolution of Victorian Science

1840–1910

Lee T. Macdonald

UNIVERSITY OF PITTSBURGH PRESS

PUBLISHED BY THE UNIVERSITY OF PITTSBURGH PRESS, PITTSBURGH, PA., 15260

Copyright © 2018, University of Pittsburgh Press

All rights reserved

Manufactured in the United States of America

Printed on acid-free paper

10 9 8 7 6 5 4 3 2 1

Cataloging-in-Publication data is available from the Library of Congress

ISBN 13: 978-0-8229-4526-0

JACKET ART: Charles Joseph Hullmandel (1789–1850) after George Ernest
Papendiek (1788–1835), "Observatory, Richmond Gardens" from *Kew Gardens*
(London: R. Ackerman, 1820), hand-colored lithograph,
Yale Center for British Art, Paul Mellon Collection

JACKET DESIGN: Joel W. Coggins

CONTENTS

CONTENTS

ACKNOWLEDGMENTS

FIRST OF ALL, I WOULD LIKE TO THANK PROFESSOR GRAEME GOODAY and the School of Philosophy, Religion and History of Science at the University of Leeds, UK, for originally encouraging me to write about the history of Kew Observatory. Graeme patiently read my work, encouraged me to think critically, and above all showed enormous dedication and professionalism. Similarly, I thank the series editor, Professor Bernard Lightman, together with Abby Collier and Alex Wolfe at the University of Pittsburgh Press, for seeing the work through to publication.

The School of Philosophy, Religion and History of Science at Leeds proved to be a stimulating environment in which to work and I benefited greatly from discussions with colleagues at seminars and lunchtime gatherings—notably Jon Topham, Adrian Wilson, Liz Bruton, and Anne Hanley, as well as Graeme Gooday. Interactions with several of these colleagues have encouraged me to think about continuities between different periods of history. In addition, teaching undergraduate courses at Leeds had the effect of broadening my perspective on the history of science.

A grant from the Arts and Humanities Research Council project, "Constructing Scientific Communities: Citizen Science in the 19th and 21st Centuries," supported the work of completing the book, and I thank the project's leader, Professor Sally Shuttleworth at the University of Oxford, for facilitating this.

My research for this book has made use of a very large quantity of unpublished archival material, spread over locations spanning nearly six degrees of latitude across the British Isles—from Exeter in the southwest of England to St Andrews on the North Sea coast of Scotland. Mark Beswick and his colleagues at the National Meteorological Archive in Exeter were ever helpful in granting me access to documents about Kew Observatory, including the unpublished minutes of the Kew Committee and the 1850–1851 "Kew Diary." Similarly, thanks are due to the National Archives for access to their Kew Observatory papers, as well as some invaluable correspondence of John Herschel and Edward Sabine. Keith Moore, Rupert Baker, and colleagues at the Royal Society Library and Archives went above and beyond the call of duty, especially with lugging out volume after volume of Herschel and Sabine correspondence. I am grateful to Jon Cable and his colleagues at the Institution of Engineering and Technology for allowing me access to the papers of Sir Francis Ronalds, which helped me to gain new insights into the early years of Kew Observatory under the British Association for the Advancement of Science. Beverley Ronalds was kind enough to provide an autobiographical letter by Francis Ronalds that she had transcribed.

The minutes of the British Association Council are housed at the Bodleian Library in Oxford and I thank the staff at the Bodleian for access to these. At Cambridge University Library, Adam Perkins and his team provided generous help with access to papers in the Royal Greenwich Observatory archives. The correspondence of James David Forbes at St Andrews University helped shed some important new light on Kew Observatory in the mid-nineteenth century and I must thank Dr Isobel Falconer for pointing me to the location of these papers. Thanks are also due to staff at the Special Collections department of Leeds University Library for access to various items and the Harry Ransom Humanities Center at the University of Texas, USA, for arranging to send electronic scans of Herschel correspondence.

Katherine Marshall and Keith Moore of the Royal Society generously gave of their time in sourcing some of the photographs that illustrate this volume, and I also thank these individuals and the Royal Society for permission to use them here. I owe a similar debt of gratitude to Mark Beswick and colleagues at the National Meteorological Archive for finding illustrations and permitting their use, as I do to Sarah Clack of the National Physical Laboratory for the image shown in figure 6.1.

More recently, I have gained much from interactions with colleagues at the National Maritime Museum in Greenwich and the Museum of the History of Science in Oxford. Here I must thank Louise Devoy and Richard Dunn at the National Maritime Museum and Silke Ackermann, Stephen Johnston, Lucy Blaxland, Robyn Haggard, Keiko Ikeuchi, and Tony Simcock at the Museum of the History of Science. Contact with scholars in the wider world of the history of astronomy, meteorology, and geophysics has also sharpened my thinking. Notable among these have been Allan Chapman, Gregory Good, Roger Hutchins, Jack Morrell, and the late Malcolm Walker. David Willis and Matthew Wild of the

Rutherford Appleton Laboratory, and also Graham Appleby of the Natural Environment Research Council, gave valuable insights into the history of twentieth-century solar-terrestrial physics.

I have presented papers based on parts of this book at various conferences and seminars, including the British Society for the History of Science annual conferences (St Andrews, July 2014 and Swansea, July 2015); the Institute of Physics conference on the history of physics (Cambridge, September 2014); the Science, Medicine and Culture in the Nineteenth Century seminar series (Oxford, May 2015); the Twelfth Biennial History of Astronomy Workshop at the University of Notre Dame, Indiana, USA, in June 2015; and the Third Biennial Early-Career Conference for Historians of the Physical Sciences, held at Annapolis, Maryland, USA, in April 2016. Here I must formally record my thanks to the School of Philosophy, Religion and History of Science at Leeds for the award of a grant that enabled me to attend the Notre Dame conference and also to the American Institute of Physics for financial support to attend the Annapolis event.

Last, but by no means least, I must thank my parents, Joyce and Raymond Macdonald, for all their support and encouragement in so many ways throughout my years of work on this book.

ABBREVIATIONS

BAAS   British Association for the Advancement of Science

BAAS Report   Report of the British Association for the Advancement of Science (see the references for reports for individual years)

BAAS:CM   British Association for the Advancement of Science, Minutes of Council. Bodleian Library, University of Oxford, UK

IET S.C.Mss.1   Institution of Engineering and Technology, London, UK: Papers of Sir Francis Ronalds, Journal of copy letters and diary entries

KCM   Kew Committee Minutes. Meteorological Office, National Meteorological Archive, Exeter, UK

KCR   Report of the Kew Committee

KD   "Kew Diary," 27 August 1850–31 October 1851. Meteorological Office, National Meteorological Archive, Exeter, UK

MMC   Minutes of the Meteorological Committee (1867–1877), Meteorological Council (1877–1905), and Meteorological Committee (1905–1910). The National Archives, Kew, UK, TNA:BJ 8

NPL   National Physical Laboratory

NPL Report   *National Physical Laboratory Report* (see the references for reports for individual years)

NPL:ECM   National Physical Laboratory: Minutes of Executive Committee

OED	*Oxford English Dictionary*
RAS	Royal Astronomical Society
RGO	Royal Greenwich Observatory Archives, Cambridge University Library, Cambridge, UK
RS:CM	Royal Society: Minutes of Council (printed)
RS:CMB.284	Royal Society: Minutes of Committee of Physics and Meteorology, 1839–1845
RS:EC	Royal Society: Election Certificate
RS:HS	Royal Society: Correspondence of Sir John Herschel
RS:MC	Royal Society: Miscellaneous Correspondence
RS:MM	Royal Society: Miscellaneous Manuscripts
RS:MS.538	Royal Society: National Physical Laboratory Papers
RS:MS.843	Royal Society: Secretarial Correspondence, 1854–1871
RS:Sa	Royal Society: Correspondence of Sir Edward Sabine
TNA:BJ 1	The National Archives, Kew, UK: Kew Observatory Papers
TNA:BJ 3	The National Archives, Kew, UK: Papers of Sir Edward Sabine
TxU	Harry Ransom Humanities Center, University of Texas at Austin, USA: Correspondence of Sir John Herschel

KEW OBSERVATORY AND THE EVOLUTION OF VICTORIAN SCIENCE
1840–1910

INTRODUCTION

Kew Observatory, Victorian Science, and the "Observatory Sciences"

One more recent instance of the operations of this Society in this respect I may mention, in addition to those I have slightly enumerated. . . . I mean the important accession to the means of this Society of a fixed position, a place for deposit, regulation, and comparison of instruments, and for many more purposes than I could name, perhaps even more than are yet contemplated, in the Observatory at Kew.

ADDRESS BY LORD FRANCIS EGERTON TO BRITISH
ASSOCIATION FOR THE ADVANCEMENT OF SCIENCE,
JUNE 1842

WHEN IN 1842 LORD EGERTON, PRESIDENT OF THE BRITISH ASSOCIATION for the Advancement of Science (BAAS), announced the association's acquisition of Kew Observatory (figure I.1), he heralded the inauguration of what would become one of the major institutions of nineteenth-century British—indeed international—science. Originally built as a private observatory for King George III and long in a moribund state, after 1842 the Kew building would, as Egerton predicted, become a multifunctional observatory, put to more purposes than were even imagined in 1842. It became distinguished in several sciences: geomagnetism, meteorology, solar astronomy, and standardization—the latter term being used in this book to refer to testing scientific instruments and developing prototypes of instruments to be used elsewhere, as well as establishing and refining constants and standards of measurement. Many of the major figures in the physical sciences of the nineteenth century were in some way involved with Kew Observatory. For the first twenty months of the twentieth century, Kew was the site of the National Physical Labora-

FIGURE 1.1. Kew Observatory in 2012. Photograph by Lee Macdonald.

tory (NPL) before the new organization moved to its present location at Teddington.

For all that, little has been written about Kew Observatory—indeed, hitherto there has been no book-length work at all devoted to its history. Part of the problem with the historiography of Kew Observatory is that Kew has always meant different things to different people. To the astronomer, it is the place where Warren De La Rue began the first systematic effort to photograph the sun. To the geophysicist, it is associated with Edward Sabine and his projects to map Earth's magnetic field. To the meteorologist, it is an almost holy place, where new types of equipment were trialed and innovations in meteorological observation pioneered. The building remained in use as a meteorological observatory until 1980, enabling some meteorologists in modern times to look back on it with nostalgia because they themselves worked there while students or trainees.[1] Additionally, science historians sometimes cite it as a "public observatory" where a new type of experimental astronomy was pioneered or as a site where data was collected in the hopes of refuting Victorian materialist cosmologies.[2] The most extensive—and widely cited—general history of Kew Observatory is the 1885 paper by Robert Henry Scott, then director of the Meteorological Office. This is a very basic, uncritical chronology that offers no analysis, or even mention, of many of the politics behind the various changes in the running of Kew Observatory in the nineteenth century. Furthermore, it is a contemporary account by a practitioner of science and so lacks the historian's perspective. Other than Scott's account, there is only a small handful of articles dedicated to Kew Observatory, all of them from a similar uncritical, internal perspective.[3]

For this reason alone, given its importance as a nineteenth-century scientific institution, this book is intended to fill a major gap in the literature on Victorian science. The history of Kew Observatory also allows us to tackle some major issues that are of great current interest to historians of science in the nineteenth and early twentieth centuries. The life of the observatory from its acquisition by the BAAS in 1842 to its becoming a purely meteorological institution in the period 1910–1914 covers a period of history from the first years of Queen Victoria's reign to the eve of the First World War—practically the entire span of what we might call "Victorian science." In this book, I will address these issues by asking three major questions about Kew Observatory.

1. WHAT CAN THE HISTORY OF KEW OBSERVATORY TELL US ABOUT HOW THE PHYSICAL SCIENCES WERE ORGANIZED IN THE NINETEENTH AND EARLY TWENTIETH CENTURIES?

The issue of the organization of the physical sciences can be divided into three subquestions. First, how were the physical sciences funded? Secondly, how were they managed? Finally, what kind of people worked in these sciences? The patronage of science and to what extent this changed over the Victorian period has long featured prominently in the secondary literature. Kew offers a good case study that can further develop our knowledge as to how patronage worked in the physical sciences, particularly as Kew is not easy to categorize: it was not a publicly supported observatory, like Greenwich, nor was it a private observatory of the kind that belonged to one of the wealthy devotees of science who played a leading role in Victorian scientific discovery. Kew can also tell us much about which individuals and organizations managed science. In particular, it has the potential to throw new light on the nature of the ancient Royal Society (founded in 1660) and the much newer BAAS (founded in 1831), as well as their relationships with each other, since both organizations were heavily involved with Kew.

Finally, a study of Kew Observatory can offer much insight into who was involved in the physical sciences. The historian David Philip Miller has identified three groups of practitioners in the physical sciences that came to prominence in Britain in the decades after the Napoleonic Wars: the "mathematical practitioners," the "Cambridge network," and the "scientific servicemen." The mathematical practitioners worked in the military colleges, such as the Royal Military Academy at Woolwich, or else they came from commercial backgrounds, often in the city of London, and put the skills they had learned in professional life to use in mathematical sciences such as astronomy. The Cambridge network comprised those who had studied for the Cambridge mathematical tripos following the reforms in the mathematics syllabus in the 1810s and had remained close friends throughout their careers. John Herschel, George Airy, and Charles Babbage can all be considered members of the Cambridge network. The scientific servicemen were army and naval officers employed in scientific surveys and other projects, especially after the end of the Napoleonic Wars freed up some military resources.[4] Other historians have identified a fourth group: physicists based in the new research and teaching laboratories that emerged later in the nineteenth century.[5]

I have also adopted the term "gentlemen scientists" to describe the many wealthy devotees of science who had time and leisure to pursue their own research interests. For the early part of the 1840–1910 period, I use the term *devotees of science* instead of *gentlemen scientists*. The word *scientist*, although coined by William Whewell in the 1830s, did not come into common use until later in the nineteenth century.[6] *Gentlemen scientists*, however, makes a useful contrast with the *university physicists* who emerged later in the century. Some of these also belong to the other categories: for example, Miller classes the stockbroker-turned-astronomer Francis Baily as one of the mathematical practitioners, though he can also be considered a gentleman scientist in that he funded his astronomical work with his private fortune. All these groups had much involvement with Kew. Overall, the story of Kew Observatory between 1840 and 1910 may help shed light on the question as to what extent the social organization of the physical sciences changed over this period.

2. HOW DID THE "OBSERVATORY SCIENCES" AT KEW DEVELOP BETWEEN 1840 AND 1910?

In the popular imagination—and even in some scholarly histories of science—an observatory is typically seen as a place devoted solely to astronomical observation. Until recently, most of the literature on the history of observatories concentrated mainly on astronomy. Yet at most observatories in the nineteenth century—especially national observatories founded by the state—those who worked in them did other sciences as well, notably meteorology, geomagnetic observations, and standardization work, such as testing chronometers for their countries' navies and merchant shipping. Some historians, notably David Aubin, are now starting to address this overwhelming dominance of astronomy in the historiography of observatories—especially with the development of the concept of *observatory sciences*, defined as sciences involving observation such as meteorology as well as astronomy that are practiced within the common space of the observatory and share the same set of techniques.[7] Aubin has argued that the nineteenth century was a time of triumph yet also of crisis for the observatory, as these rapidly developing institutions had to adapt in order to accommodate new fields of work and communicate the results of that work through new media such as photography and the electric telegraph. Meteorology, for example, became a central part of the program of work at many observatories, including Kew; the results of meteorological observations were communicated and coordinated via the expanding telegraph network. Yet by the end of that century, the situation had changed again: observatories tended to specialize in just one observatory science, while each of the observatory sciences had come to be managed by separate, specialized institutions of state.[8]

Kew offers a better case study than most observatories with which to trace the evolution of the observatory sciences because a wider variety of these sciences was practiced at Kew than at most observatories of its time. In fact, after 1842, astronomy was not Kew's main purpose but just one of a diverse range of activities there. Kew became a national nerve center for several sciences that today are administered by five separate institutions: meteorology (now under the Meteorological Office), solar physics (run by the Science and Technology Facilities Council), standardization (National Physical Laboratory), and geomagnetism (British Geological Survey and Natural Environment Research Council). Yet Kew Observatory became less of a nerve center and more specialized as the century drew to a close: in meteorology it became an outstation, reporting to the Meteorological Office in London; solar astronomy moved to Greenwich; geomagnetism became predominantly routine work under the control of university physicists; while standardization emerged as the most important activity at Kew. Although it was still called "Kew Observatory" in the late 1890s, by then it was primarily a standardization laboratory: only a small portion of the work carried out there involved observation of external phenomena. The balance of activities at Kew had shifted from the work of an observatory toward that of a laboratory. Then, after the 1910 transfer of responsibility for Kew from the NPL to the Meteorological Office, the observatory came to specialize in just one science, meteorology. In Kew Observatory, we have a case study in the process of specialization in the observatory sciences during the late nineteenth and early twentieth centuries. It allows us to study the history of how these sciences evolved over the course of this period—with the added benefit that we do not have to take into account the many variables involved when considering the history of more than one institution in more than one country.

3. HOW DID STANDARDIZATION DEVELOP AT KEW IN THE CONTEXT OF THE CULTURE OF THE PHYSICAL SCIENCES BETWEEN 1840 AND 1910?

Various historians have described how the establishment of agreed physical constants and standards of measurement, including the precision instruments needed for making the measurements, became an important part of the culture of the physical sciences in the early nineteenth century. Standardization became an essential component of nation building, notably in revolutionary and postrevolutionary France, the German states during the same era, and in Britain after the end of the Napoleonic Wars. In the British case, there was a need for comparability of weights and measures across a large empire, for commercial and legal as well as scientific purposes. Government and business alike wanted reliable standards of length and weight to maintain Britain's preeminent position in global trade and also to reduce the widespread fraud that was allegedly encouraged by long-standing regional variations in British weights and measures. In 1824, an Imperial Weights and Measures Act had finally established a system of standards of length and weight, enshrined in law, after centuries of failed legislation.[9]

The British government's increasing interest in standardization began to affect Kew in the early 1850s when some army officers working in the Indian subcontinent became interested in meteorology, with a view to governing this part of the empire more efficiently using an improved knowledge of the weather. This led to the British East India Company wanting thermometers and other instruments that would be comparable with each other across the imperial possessions in India and beyond. Then, in 1853, United States Naval Observatory superintendent Matthew Fontaine Maury made moves to extend his system of ocean weather charts to all oceans around the globe. For this to become a reality, it was necessary to institute an internationally accepted system for recording weather observations aboard ships. Such a system was agreed to at an "International Meteorological Conference," held in August and early September 1853. Ten nations were signees, including Britain. To issue naval and merchant shipping with standardized meteorological instruments, as well as to administer the collation of the weather data obtained, the British government set up a new department, known initially as the Meteorological Department of the Board of Trade and headed by Robert FitzRoy, formerly captain of HMS *Beagle* during Charles Darwin's voyage around the globe.[10] Starting from the mid-1850s, the instruments issued by the Meteorological Department would all be tested at Kew Observatory, which rapidly became the preeminent place in Britain for testing meteorological and also magnetic instruments—not only for British ships and imperial observatories but for institutions in other countries as well. Yet existing accounts of nineteenth-century meteorology take this role of Kew Observatory for granted. There have been two brief descriptions of instrument testing at Kew,[11] but there has been no discussion as to how and why instrument standardization began there in the early 1850s.

In the second half of the nineteenth century, the growth of research and teaching laboratories in universities led to an expansion of the demand for precision measuring instruments. Precision measurement in the nineteenth-century laboratory has been well covered in modern scholarship with respect to universities—in particular, the rising generation of university physicists. In six case studies of university laboratories, Graeme Gooday has shown how these teaching laboratories trained undergraduates in the skills of laboratory measurement that were essential to the training of—and satisfying the growing demand for—school science teachers and entrants to the burgeoning electrical engineering profession. Similarly, Simon Schaffer has described the rise of measurement science at the Cavendish Laboratory in Cambridge and its relation to industry.[12] However, what of the institution that provided the precision instruments that were so essential not only to the university teaching laboratories but also to institutions such as the Meteorological Office, the Admiralty, and the merchant marine? Before the NPL opened in 1900, and in some cases even afterwards up to 1910, that institution was Kew Observatory. Again, practically nothing has been written on standardization at Kew in the histories of the NPL or in the wider literature on Victorian physics in relation to either the world of late Victorian science and industry or on its role in the origins of the NPL. One of the aims of this book is to address this gap in the literature.

LAISSEZ-FAIRE AND THE PHYSICAL SCIENCES, 1840–1910

Historians generally agree that the governance of science in Britain underwent profound changes in the decades between 1840 and 1910. Since the 1820s, attempts had been made to reform the Royal Society, Britain's most prestigious scientific body and the one with the most influence over government, from what some perceived as a club for wealthy gentlemen into a learned body representing the most serious and able practitioners of science. This change did not come easily: only in 1847 was the society's constitution amended so that admission to fellowship was granted on scientific merit alone.[13] Long after 1847, the issue of who should run the Royal Society was sometimes a contentious one.[14] In the meantime, the British Association for the Advancement of Science had emerged as a rival organization. The BAAS was founded in 1831 after a failed attempt by some leading men of science to reform the Royal Society. It had a much more democratic structure than the older body in that all decisions taken by its council had to be ratified at the association's annual meetings, which were held in a different provincial town each year, a deliberate break away from the Royal Society's image of an exclusive London club. Yet the distinction between the two bodies was not as clear-cut as might at first appear. In the absence of regular government grants, the BAAS still needed the support of wealthy aristocrats in order to gain influence and money.[15] In practice, many leading men of science, whatever their social position, belonged to both organizations—something that would have a strong influence on the development of Kew Observatory at various times in its history after 1842.

Systematic government grants to the Royal Society only commenced in 1850.[16] Before then, with the exception of scientific organizations connected with the army and navy, government financial support for science was on a strictly ad hoc basis, gained largely through lobbying and persuasion by grand figures, usually via the Royal Society. Failing this, funding had to come from private individuals or, after 1831, from the BAAS's limited funds, which originated largely from members' subscriptions and private donations. Significantly for the history of Kew Observatory, the BAAS's usual policy was to fund individual projects of limited duration or perhaps make grants to allow the purchase of equipment for specific purposes but not to support permanent scientific programs or institutions.[17] Therefore the attempts to gain financial support to transform Kew Observatory in the early 1840s—and to keep it running later that same decade when it was threatened with closure—can tell us much about the sources of patronage that devotees of science had to find in the second quarter of the nineteenth century, before government money became available on a regular basis.

There is broad agreement among scholars that in the first two-thirds of the nineteenth century, science-government relations followed the prevailing economic consensus of laissez-faire—the doctrine that government should not interfere in an economy that was presumed to be self-regulating.[18] The £1,000 government grant given to the Royal Society each year from 1850, if anything, encouraged this system: individuals could apply for money out of this grant to buy equipment for their own research, and so it rewarded individual enterprise. The grant was never intended to fund salaries or long-term projects. This situation began to be challenged in the late 1860s. At the BAAS's 1868 annual meeting, Lieutenant-Colonel Alexander Strange, a former inspector of scientific instruments for the Indian trigonometric survey, gave a paper whose very title expressed his views in one sentence: "On the Necessity for State Intervention to Secure the Progress of Physical Science."[19] Strange believed that the government had to invest more money in scientific education and research institutions if it was to keep up with increasing overseas competition in science and—particularly close to the heart of this former army officer—govern the British Empire effectively. Prominent among those agreeing with Strange was Lyon Playfair, who had helped to organize the 1851 Great Exhibition at South Kensington and who had afterwards campaigned for greater government input into science education. Both Playfair and Strange had served as jurors in the 1867 International Exposition in Paris, after which Playfair famously expressed alarm at how far foreign inventions had caught up with Britain since the 1851 exhibition.[20]

Strange's views caught on at the BAAS and his paper was enthusiastically taken up by some senior BAAS members and Fellows of the Royal Society. This led to a successful lobby for a Royal Commission to look into the state of science education and—most importantly for the history of Kew Observatory—that of institutions for scientific research. The commission, which ran from 1870 to 1875, was chaired by William Cavendish, seventh Duke of Devonshire (himself a Cambridge mathematics Wrangler) and hence became known as the Devonshire Commission. Its final report, published in 1875, recommended the establishment of more government-funded laboratories, including a new observatory dedicated to the physics of astronomy.[21]

Some well-known twentieth-century works on the organization and funding of science see the period of the Devonshire Commission as representing the start of organized science—meaning professional scientists working in government institutions or large companies, in contrast to the earlier regime in which science had largely been carried out by wealthy individuals working on their own time. For Donald Cardwell, in particular, there was no such thing as "the social organization of science" before the mid-nineteenth century. The history of British science before then was just a "preface" to it: "Important and not without historical interest, but still a preface."[22] Authors of Cardwell's generation all wrote during the third quarter of the twentieth century, at a time when science in Britain enjoyed generous state support and there was widespread agreement that it *should* enjoy such support. This led many historians to take a teleological view, seeing large-scale government investment in research institutions as inevitable—something since admitted by Roy MacLeod.[23] These authors generally concede that the initial impact of the Devonshire Commission on governments was small and that only slowly were its recommendations taken up. Yet they treat the end of the nineteenth century as a time in which twentieth-century state-supported science finally began to triumph over nineteenth-century laissez-faire —as symbolized by the establishment in 1900 of the National Physical Laboratory, an institution founded as a British answer to Germany's generously state-funded Physikalisch-Technische Reichsanstalt.[24]

Kew Observatory, however, does not fit into this tidy picture. In addressing the issue of the organization of science, one of the aims of this book is to use the history of Kew Observatory to challenge the idea that laissez-faire—and the physical sciences' consequent reliance on private sources of patronage—went out of fashion before the end of the nineteenth century. For in the chapters that follow I show how right up until it became part of the NPL in 1900, Kew remained an exemplar of the laissez-faire system in action. Before 1900, it received relatively little money from government grants. Most of its work was funded from private sources and—increasingly important later in the century—from the fees it charged for testing instruments on behalf of manufacturers and government bodies. In particular, I contend that the birth of the NPL was facilitated not by a change in the government's attitude but rather by the sheer lack of government support for observatories and laboratories. As the nineteenth century drew to a close, the ever-pressing need to make money forced Kew to turn itself effectively into a national standardization laboratory and so form the basis of a ready-made NPL. Historians of the NPL have shown that even after 1900 it retained many of the characteristics of Kew Observatory in the nineteenth century: its leaders continued to grumble about lack of funding, and the Treasury expected it to be self-supporting through the fees it charged for instrument tests.[25] In chapter 6 of this book, I argue that laissez-faire economics had an important bearing on the development of Kew Observatory, now the "Observatory Department" of the NPL, between 1900 and the eve of the First World War. This book thus challenges and revises the view of Cardwell and others as to the demise of laissez-faire with regard to scientific funding in the late nineteenth century. Rather, it aims to present a sense of continuity between Kew Observatory and the NPL and hence to show that in government support for the physical sciences, laissez-faire remained predominant into the first years of the twentieth century.

OBSERVATORIES IN NINETEENTH-CENTURY BRITAIN

Susan Faye Cannon and David Philip Miller have both pointed to the three decades following the end of the Napoleonic Wars as a period of expansion and increased cooperation in the physical and mathematical sciences.[26] A notable feature of this movement was the construction of many new observatories and the adaptation of older ones to new purposes, among them nonastronomical sciences. Dieter Herrmann has shown how the establishment of new observatories worldwide increased exponentially during the nineteenth century, from 31 in 1810 to 199 in 1900. It was also during the nineteenth century that the word *observatory* became common in English literature—and therefore culturally significant—as David Aubin has demonstrated using Google Books.[27]

Observatories in the nineteenth century can be grouped into three broad categories: national, university, and private observatories. This book uses the phrase *national observatory* as defined by Steven Dick: an observatory established, funded, and staffed by a national government for a purpose that the government deemed to be of national importance at the time. Dick has suggested that the nineteenth century saw the second wave of an overall "movement" to build national observatories; this movement began in the sixteenth century and was still continuing in the late twentieth century.[28] National observatories were founded by governments for very specific purposes, of which the main one was usually the measurement of celestial positions to supply data for navigation. The most prestigious national observatory in Britain was the Royal Observatory at Greenwich. Founded in 1675 to solve the problem of finding longitude at sea, by the 1830s Greenwich was a world standard in navigational astronomy. The observatory provided data for the production of tables providing stellar, planetary, and lunar coordinates that enabled sailors to find their position at sea quickly and accurately.

By the early 1830s, however, Greenwich was in some disarray. The reductions of observations into a form usable for longitude tables had fallen into arrears, and relations between Astronomer Royal John Pond and his staff were poor. In 1835 the Admiralty, the government department responsible for the Royal Observatory, replaced Pond with the thirty-four-year-old Cambridge mathematician George Biddell Airy. As Robert Smith and others have shown, Airy quickly turned the Royal Observatory into a factory-like regime that efficiently produced quality data for navigation and, later on, a national time service. Airy had such a powerful influence over the Royal Observatory that his name is practically synonymous with Greenwich between the mid-1830s and the early 1880s, despite his occasional disagreements with James Glaisher, the head of the magnetic and meteorological department at Greenwich from 1840. Airy saw himself primarily as a public servant. He believed that research with no immediate utilitarian purpose, such as sweeping the heavens for new nebulae or planets, lay outside the remit of Greenwich and should be left to private or university observatories that did not spend the state's money.[29] Yet he did not take kindly to criticism, nor to incursions by other public institutions onto territory that he felt was his. This would have an important bearing on the history of Kew Observatory from the 1840s onward.

Roger Hutchins has described how six observatories were established at universities in Britain and Ireland between the late eighteenth and early twentieth centuries. Their principal purpose was to facilitate undergraduate teaching in astronomy. In theory, they also worked in nonutilitarian branches of astronomy, such as stellar cataloging, measurements of double stars, and observations of comets, but in practice the demands of teaching often left little time for such work.[30] Forming a third category of observatories were the private observatories owned by wealthy devotees of science who spent their own money on astronomy. These gentlemen scientists were free to pursue their own agendas, as they were not required to teach or to do utilitarian work for the state.[31] Private observatories were not new in the nineteenth century, but many more of them were built after 1800. In the most comprehensive general survey of nineteenth-century amateur astronomy, Allan Chapman notes that in 1884, Armagh Observatory director John Louis Emil Dreyer published a list of some twenty-six private observatories that had done important work in the United Kingdom over the previous one hundred years.[32] A fourth category, "public observatory," has also been suggested, meaning an observatory owned and operated by a public body, such as a learned society or local government. Kew Observatory in the nineteenth century, run by the BAAS and then the Royal Society, has been described as a "public observatory."[33] However, the narrative of Kew as related in this book shows that "public" is not an easy category to apply to observatories.

Until the early nineteenth century, all three types of observatories concentrated more or less entirely on astronomy—and mostly one type at that: the "classical" astronomy of positional measurement.[34] This was dictated by the need of national observatories to serve the state, but the other types of observatories tended to concentrate on classical astronomy too, partly because before the advent of photography and spectroscopy, it was difficult to find out anything new about the physical nature of astronomical objects. The research on nebulae by the Herschels and Lord Rosse was an exception to this general rule. Then, in the 1830s, some observatories, including Greenwich, began serious work in two sciences that hitherto had not necessarily formed part of their routine—or, at most, had been incidental to that routine: geomagnetism and meteorology. At the beginning of the nineteenth century, geomagnetism and meteorology hardly existed as sciences organized on a national scale. In Britain, this situation persisted into the early 1830s, with geomagnetic work being done by isolated individuals such as the Royal Artillery officer Edward Sabine at Woolwich.[35] Elementary meteorological observations were being carried out at a small handful of locations, such as Kew Observatory (then still known as the "King's Observatory"), the Royal Society's headquarters at Somerset House, and the Radcliffe Observatory at Oxford, as well as by a few private individuals, but the science was not organized on a national scale until the 1850s.[36] But when these two observatory sciences did take off, they did so together. They were seen as being closely connected, for several reasons. Many thought that changes in the weather and Earth's magnetic field were subtly related to each other, or that both had astronomical origins, and in any case temperature and pressure were found to affect magnetic compass readings. Both sciences had clear importance to navigation in an age when Britain was the world's chief maritime power. In particular, the reasons for the behavior of the compass aboard ships were poorly understood, as were the weather and currents in many parts of the oceans. It was in this context that in the 1830s, some observatories began making systematic meteorological observations and also began monitoring Earth's magnetic field as part of a global campaign known as the "Magnetic Crusade," described in chapter 1.

THE ORIGINS AND EARLY HISTORY OF THE "KING'S OBSERVATORY"

The origins of Kew Observatory are well known and well documented. Nineteenth-century sources agree that it was originally known as the "King's Observatory"; it came to be called the "Kew Observatory" some years prior to 1840.[37] In an 1839 letter to John Herschel, Admiralty hydrographer Francis Beaufort remarked, "Perhaps I should have called it the Kew Observatory"—suggesting that the building had only recently come to be known by this name.[38] It was built in 1768–1769 for King George III to enable him to observe the transit of Venus on 3 June 1769. The building was designed by the eminent architect Sir William Chambers (who went on to design Somerset House) and was completed in time for the transit, which was successfully observed by the King and others in a clear sky.[39]

However, this spectacular beginning to the observatory's career was not matched by the work done in the years that followed, for it was not used, nor even intended for, astronomical research or navigational astronomy of the kind being done at Greenwich. To run the observatory the King appointed his former tutor, Stephen Charles Triboudet Demainbray, a much-traveled university lecturer of French Huguenot descent, as his "King's Observer." After the transit (which Demainbray observed with the King), Demainbray's duties seem to have been light. His principal duty was to take daily transit timings of the sun as it crossed the meridian; these observations were used to regulate high-quality clocks that kept standard time in the observatory and at several prestigious public places in London, among them the houses of Parliament. Basic meteorological observations, including recordings of temperature and rainfall, were commenced in 1773 and continued until 1840, with the thermometers placed in a north-facing window and the rain gauge mounted on the roof. The observatory was also used as an instrument repository and a place where members of the royal family received tuition from Demainbray. Kew was included in a 1777 survey of observatories by Copenhagen Observatory director Thomas Bugge, who noted that the building contained numerous instruments, including a transit telescope and a large mural quadrant. Bugge also noted that the basement contained "mathematical workshops."[40] When Demainbray died in 1782, the King appointed Demainbray's son, the Reverend Stephen Demainbray, as his observer at Kew. Both Demainbrays were assisted in the observations by their fellow Huguenot and family relative Stephen Peter Rigaud. Upon Rigaud's death in 1814, the job of assistant went to his son, also called Stephen Rigaud. Rigaud Jr. had been Savilian Professor of Geometry at Oxford since 1810 before he became Savilian Professor of Astronomy at the same university in 1827. He took over the running of Kew Observatory during the university's summer vacations, thus allowing the Reverend Demainbray to live in his Wiltshire parish during the summer months. Demainbray, Rigaud, an assistant, and a servant all appear to have drawn salaries for their work at Kew.[41] In 1827 Rigaud's wife died, leaving him to bring up his children on his own as well as perform his academic duties at Oxford. Although still officially an observer at Kew, he was seldom able to go there from then on. By this time, too, George III was dead and his successors to the throne took less interest in the observatory. This, plus the observatory's substantial salary costs, might well have been a motive for the government to drop its support for Kew.

It is easy to think of these shared jobs of the Demainbrays and Rigauds as sinecures and that the King used the building as little more than a showcase for his instrument collection. Yet Bugge's survey notes that the observatory contained some of the best equipment that money could buy at the end of the eighteenth century, including a mural quadrant and a precision measuring telescope, both made by leading astronomical instrument maker Jonathan Sisson.[42] A list of the observatory's astronomical instruments presented to Armagh Observatory in 1841, when the government withdrew its support from Kew, also includes some high-quality examples.[43] It was in the King's observatory that John Harrison's "H.5" marine chronometer was given its final successful test that enabled Harrison to claim the remainder of his share of the £20,000

"Longitude Prize." The chronometer was tested in the observatory over a ten-week period between May and July 1772. It was regularly compared with the observatory's clock, which was itself checked with meridian transits of the sun.[44] The transit timings were taken with a transit telescope suspended between two massive masonry piers on the ground floor of the observatory. This provided as good a time service as any at the end of the eighteenth century: before the advent of telegraphic communications, Greenwich was remote from Kew and central London, so time had to be determined and distributed locally.[45] Bugge noted that the foundations of the building "were laid 20 to 30 feet below the ground" in order to ensure a stable platform for the astronomical instruments.[46] In 1843, soon after becoming honorary superintendent at Kew under the BAAS, Francis Ronalds would make a remark that corresponded exactly with Bugge's notation: that the building's foundation was "of an extremely solid and costly kind."[47] In the mid-1840s, Ronalds would adapt the transit pillars to another type of precision measurement: the monitoring of tiny variations in Earth's magnetic field using a magnetometer suspended between these pillars.

Thus in Kew Observatory, the BAAS and the Royal Society had a ready-made space for precision measurement; it is clear from the evidence just noted that Ronalds was well aware of this. The building's suitability for precision measurement would have an important bearing on its history after 1842. Some modern scholarship has discussed how buildings such as the Physikalisch-Technische Reichsanstalt in Berlin were deliberately designed and built with metrology in mind.[48] Kew provides an opportunity to see how an existing building, constructed for astronomical and meteorological observations in an earlier age, was adapted for the measurement sciences of a later era.

PRIMARY SOURCES AND THE SCOPE OF THE BOOK

The volume of primary-source material on Kew Observatory increases as we progress through the nineteenth century. It is possible to learn much even from published primary sources, as few of them have ever been cited by historians. Reports of the Kew Committee appear regularly in the *BAAS Annual Report* from 1850 until 1871 inclusive; thereafter they can be found each year in the *Proceedings of the Royal Society*. These reports run to several (latterly over twenty) pages each and describe the previous year's activities at Kew in some detail. From the late 1850s they contain detailed financial accounts, including lists of the observatory's employees and their salaries. The volumes of the *BAAS Report* also contain many papers on specific projects at Kew, as do the Royal Society's *Proceedings* and *Philosophical Transactions*. But the value to the historian of these published sources is limited by their containing only what the members of the Kew Committee wanted their readers to hear. Like Scott's 1885 history (which is largely based on the annual reports), they frequently gloss over key developments, such as how and why John Peter Gassiot set up the trust that enabled the Royal Society to take over the running of Kew in 1871. Furthermore, very little primary-source material has been published at all on Kew before 1850. Therefore, to build a fuller picture of what happened at Kew in the period under discussion, it is necessary to turn to unpublished sources.

A large amount of archival material has survived, in the form of voluminous correspondence and minutes of meetings. The most important sets of minutes for the historian of Kew Observatory are those of the BAAS Council and the Kew Committee. The BAAS Council minutes are essential for establishing the basic narrative of events relating to Kew Observatory before the regular publication of Kew Committee reports began in 1850, especially as the correspondence for these early years is sometimes scattered and hard to find. These minutes were printed but not published, and so were not intended for general circulation. Those at the Bodleian Library in Oxford are mostly complete to 1868; copies relating to the years from 1868 to 1871, the period leading up to the handover of Kew from the BAAS to the Royal Society, are preserved in the files of the ever-meticulous George Airy.

The Kew Committee began taking formal minutes of its meetings in October 1849, and so from this date we can assemble a more detailed narrative. The minutes of the Kew Committee were handwritten in minute books and never printed, so they contain many details of the observatory's history that were confidential at the time. Furthermore, these minutes have never been used by *any* modern scholar, enabling us to discover vast amounts of new information and gain important new perspectives. The minutes for the post-1871 period are especially useful because they frequently refer to numbered correspondence. These letters are preserved in the National Archives at Kew and many of them still bear their original index numbers, making it easy to find many letters referred to in the minutes of the Kew Committee. Minutes for the 1840–1910 period tend to record merely a summary of what was agreed upon at a meeting, rather than what was actually discussed. Like the published sources, they sometimes present only an official version of events, leaving out the arguments and disagreements.[49] Nevertheless, due to their confidential nature, they contain many telling details that have been left out of the published record of events.

The richest—and most revealing—set of unpublished sources is the correspondence of the numerous individuals who were involved with Kew Observatory. The letters of Francis Ronalds, Kew's first superintendent, provide important insights into Kew's very first years under the BAAS, especially when read in conjunction with the BAAS Council minutes. The most useful correspondence for these early years is that of John Herschel, not only because of his views on observatories and his involvement in so many of the behind-the-scenes moves regarding Kew Observatory in the 1840s, but also because of his centrality to—and per-ceived authority in—so many of the physical sciences in these years. His approximately 15,000 incoming and outgoing letters are made all the more accessible by the invaluable *Calendar* of his correspondence, which outlines the location, reference, date, and brief details of each letter.[50] This allows letters to and from Herschel referring specifically to Kew Observatory and kindred subjects to be accessed very efficiently in the Royal Society archives and elsewhere.[51] The correspondence of George Airy, held at the Royal Greenwich Observatory archives in Cambridge, is indexed online, with brief details of each file, allowing relevant letters to be accessed quickly by ordering specific files. Airy's correspondence is especially useful in that Airy kept carbon copies of his outgoing letters, enabling the historian to read Airy's replies without having to visit the papers of the people he was writing to. This is especially important in the case of the many private individuals involved with Kew whose papers are now difficult to find.

The official papers of Kew Observatory at the National Archives are voluminous and the files are indexed online, albeit with no details. Some of the files, especially from the 1870s, mostly describe trivial day-to-day matters that add little to our overall understanding. Yet we can learn much from the correspondence of John Welsh, Balfour Stewart, and the Kew Committee, most of it unread by modern scholars. The Kew Observatory papers are part of a larger collection of papers of the Meteorological Office, which includes some important correspondence of Edward Sabine. Both these Sabine papers and the Kew Observatory files at the National Archives include some letters from John Herschel that are not indexed in the Herschel *Calendar*. Easily the largest repository of Sabine's correspondence is held in the Royal Society archives. These letters are not indexed, but they are filed alphabetically by correspondent, allowing us to easily find letters to Sabine from Herschel, Gassiot, and many of the other principal actors in the history of Kew Observatory between 1840 and the early 1870s.

The total volume of correspondence, even that relating directly to Kew Observatory, runs to many thousands of letters. Only those letters most helpful to my arguments and research questions have been cited in this book. The value of such a large volume of correspondence to the historian is twofold. First, it can be used to establish an almost day-by-day chronological narrative that can give a sharper picture of the development of Kew Observatory than can ever be put together from the published sources or than has ever been attempted by historians. Secondly, it can help reveal those *unofficial* views that the actors in the story of Kew Observatory might never have wanted to reveal to many of their colleagues or the wider public—more than is often possible in minutes.

This correspondence, as well as unpublished minutes, makes it possible to challenge and revise the received views about Kew Observatory, especially in the light of modern scholarship on nineteenth-century observatories, laboratories, and physical sciences generally. This helps to tackle the three great questions about Kew outlined earlier in this chapter. To achieve these aims, the book is divided into six chapters, each covering a distinct period, in part because for each period some specific questions can be asked. The chapters are arranged chronologically, in order to show how Kew evolved over time.

Chapter 1 covers only five years (1840–1845), but this short period deserves a chapter of its own because it was in these years that the Kew Observatory of the Victorian era was founded. This chapter asks the question, How and why was the Kew building transformed from an unused royal observatory and instrument repository into what some in the BAAS called a "physical observatory"? It then asks, What work did Kew Observatory carry out during its first years? It addresses the question of the organization of science by demonstrating the importance of Edward Sabine as the prime mover behind the project to turn Kew into a magnetic and meteorological observatory and showing how he used the interchangeability between the Royal Society and the BAAS to his advantage. I show that lack of government funding did not prevent Sabine from setting up his own observatory at Kew, independent of Greenwich. I also critically assess claims that Kew was a "physical observatory" of the kind described by historians writing about the observatory sciences or of the kind advocated by Herschel.

Chapter 2 covers the period 1845–1859, from the first attempts by the BAAS to close down Kew Observatory up to the death of John Welsh, its first paid superintendent, in 1859. Here I ask how Kew withstood the moves to close it and relate this to the introduction of the Royal Society government grant in 1850. Then I chart how the observatory sciences at Kew expanded to include a full geomagnetic program as well as the meteorological work. This chapter also begins to address the third of this book's fundamental questions: How and why did standardization originate and develop at Kew? I argue that the reasons for the introduction and expansion of instrument testing at Kew were due to factors of both demand and supply. On the one hand, the government needed large numbers of thermometers, barometers, and hydrometers, all tested to an agreed standard—especially when the Meteorological Department of the Board of Trade was established in 1854. Even before 1854, however, Kew began testing instruments in return for fees because it brought in much-needed extra income.

Chapter 3 describes the period of Kew Observatory's history that has already been most discussed by historians: the pioneering program carried out in the 1860s to photograph the sun and to relate sunspot periodicities to terrestrial magnetism and weather. The narrative begins in the early 1850s, overlapping with the time span of chapter 2, in order to address the question of how and why solar photography began at Kew. I also ask how the photoheliograph was used in practice. I show how the solar photography program was largely a private enterprise, directed by gentlemen scientists and implemented by little-known figures. Finally, I explore how this new observatory science of solar physics interacted with Sabine's magnetic and meteorological agenda. I build on the existing historiography in this field to show that Stewart's conflicts with Sabine owed as much to Stewart's vastly increased workload following the Meteorological Department's reorganization as to Sabine's disagreement with Stewart's theory-driven approach.

Chapter 4, covering the years from 1871 to the publication of Robert Scott's history in 1885, asks how and why BAAS decided to stop supporting Kew and what were the circumstances surrounding Gassiot's donation that were supposed to allow the Royal Society to run it. This chapter, too, sees Airy winning a partial battle in his long rivalry with Kew: I ask why the Kew photoheliograph was transferred to Greenwich and why Airy nevertheless failed to wrest control of the Kew meteorological observations. This provides significant new insights into the changing organization and specialization of the sciences from the 1870s onward, as does this chapter's finding that by the 1880s Kew was no longer taking the lead in magnetism and meteorology; rather, its work in these observatory sciences was increasingly in the service of other organizations. I also show that by 1885, standardization had become the most important branch of the work at Kew and argue that the standardization question is intertwined with the organization of science question. Contrary to assertions that Gassiot "came to the rescue"[52] in setting up his trust to run Kew, the Gassiot fund was never sufficient to support the observatory, and the Kew Committee needed to take on more standardization work due to the money it brought in.

A central question in chapter 5 is, How and why did Kew Observatory become part of—and the first site for—the National Physical Laboratory? I show that the existence of Kew Observatory was essential to the establishment of the NPL. I also argue that contrary to assertions by historians that the NPL was a triumph of government-supported science over prevailing laissez-faire attitudes, the NPL came into existence in the form it did precisely because of laissez-faire. The continuing dominance of laissez-faire is further emphasized in chapter 6, which describes the evolution of Kew Observatory over its ten years as the "observatory department" of the NPL before it became part of the Meteorological Office in 1910. Laissez-faire remained central to government policy toward the NPL and the Meteorological Office as well as Kew Observatory. This chapter further illustrates the increasing specialization of the observatory sciences, particularly with the establishment of the new magnetic and meteorological observatory at Eskdalemuir in 1908, after which Kew effectively ceased to be the multifunction institution that it had been throughout practically the entire Victorian era and beyond.

The concluding chapter returns to the book's three overall research questions. It attempts to answer each of them using the findings presented in chapters 1–6 and thereby assesses the importance of Kew Observatory in the history of the physical sciences in the nineteenth and early twentieth centuries. The conclusions critically examine and revise some currently accepted views, especially as to the origins of the NPL and, more broadly, the evolution of the observatory sciences and their relations with government in Britain during the Victorian era.

1

A "Physical Observatory"

KEW, THE ROYAL SOCIETY, AND
THE BRITISH ASSOCIATION, 1840–1845

The observations most appropriate for the ready and exact determination of physical data are . . . those which it is most necessary to have performed with exactness and perseverance. Hence it is, that their performance, in many cases, becomes a national concern, and observatories are erected and maintained, and expeditions despatched to distant regions, at an expense which, to a superficial view, would appear most disproportioned to their objects. But it may very reasonably be asked why the direct assistance afforded by governments to the execution of continued series of observations adapted to this especial end should continue to be, as it has hitherto almost exclusively been, confined to astronomy.

JOHN FREDERICK WILLIAM HERSCHEL, 1830

Ld. Dungannon [sic] had examined the house late the Kew Observatory, and finds it in such excellent order that he will not pull it down as intended—he asked Beaufort if he knew any use that could be made of it. . . .

EDWARD SABINE, 1841

WHEN JOHN HERSCHEL WROTE *A PRELIMINARY DISCOURSE ON THE STUDY of Natural Philosophy* in 1830, he was very widely respected and arguably Britain's foremost practitioner of the physical sciences. The first epigraph above is from part 2 of the *Preliminary Discourse*, in which Herschel made a plea for state-funded observatories that collected not only astronomical data but also "physical data" such as meteorological observations and data for the determination of physical constants such as mean sea levels. In the years after 1830, John Herschel's call was enthusiastically taken up by his colleagues, and the term "physical observatory" was coined to describe the kind of observatory that Herschel wanted to see

24

established. This chapter describes how in the early 1840s the former King's Observatory at Kew was transformed into what some claimed to be such a physical observatory. How and why this happened has hitherto not been analyzed in detail. I begin by assessing the state of geomagnetism and meteorology in the early nineteenth century and how the concept of a physical observatory was developed by Herschel and others. I then use a chronological framework to show how the Royal Society, after an abortive attempt to establish a magnetic and meteorological observatory, in the end turned down the government's offer of the Kew building, which was then taken up enthusiastically by BAAS. Finally, I describe and assess the program of work implemented at Kew up to the mid-1840s. I argue that the institution that emerged was different in many ways to Herschel's vision of a physical observatory: in particular, none of the work being done at Kew up to 1845 was funded by the government.

Although *Preliminary Discourse* was clearly an inspiration behind the relaunched Kew Observatory, the story of its transformation is a complex one that owes as much to the personalities and politics of the physical sciences in the 1840s as it does to Herschel. In particular, it will become clear from this chapter that the prime mover behind the Kew project was the Royal Artillery officer Edward Sabine, one of the "scientific servicemen" identified as a group by David Philip Miller in his synoptic survey of the physical sciences in the early nineteenth century.[1] Sabine, the chief mastermind of the British geomagnetic program that came to be known as the "Magnetic Crusade," saw the building's potential as an observatory very early on—as is suggested by the second epigraph I've included. I also argue that although only a limited range of observatory sciences was practiced at Kew before 1845, the very lack of government funding in the age of laissez-faire gave Sabine a free hand to establish his own research program there, independently of his rival at Greenwich, the astronomer royal George Airy.

"PERHAPS ALL THIS IS DREAMING": MAGNETISM, METEOROLOGY, AND PHYSICAL OBSERVATORIES

Calls for improvements to the still locally organized science of meteorology were beginning to increase from the 1820s onward. However, practitioners of science realized that little progress could be made while meteorological instruments and observations remained in their existing state. In 1823, the battery and hygrometer inventor John Frederic Daniell drew attention to the poor state of the Royal Society's meteorological instruments at Somerset House and the inaccuracy of the observations made with them. Daniell was a council member of the short-lived Meteorological Society of London, which in 1824 anticipated future events by calling for accurate series of meteorological observations to be made throughout the British Empire and made comparable with each other using standardized instruments. For this to be possible, the society "should set the example of the requisite precision by establishing a Meteorological Observatory in the metropolis, or its vicinity."[2] Indeed, Daniell was later a member of the Royal Society Council and the society's Committee of Physics and Meteorology, both of which deliberated as to whether to take on Kew Observatory in the 1840s. The Meteorological Society proved to be short-lived and its proposals came to nothing, but meteorology was on the agenda of BAAS soon after its formation in 1831. The shambolic state of meteorology was stated more bluntly than Daniell by the Edinburgh natural philosopher James David Forbes in a report read to the 1832 BAAS meeting, in which he lamented that "meteorological instruments have been for the most part treated like toys, and much time and labour have been lost in making and recording observations utterly useless for any scientific purpose."[3] Forbes went further at the 1840 BAAS meeting, calling for the establishment of well-equipped "public observatories" that would "furnish standards of comparison, to establish the laws of phaenomena and to fix *secular*, or normal data."[4]

Geomagnetism began to gain prestige and public importance thanks to the well-publicized works of the Prussian explorer and scientific polymath Alexander von Humboldt. Observational work was stimulated by both Humboldt and the mathematical physicist Carl Friedrich Gauss when they began to give the subject a firm theoretical basis and demanded large quantities of accurate data with which to test their theories. They asked that this data be produced by a system of geomagnetic observatories scattered across the globe. Within a few years, such a system of observatories became a reality across the German lands and beyond, including Russia.[5] Many British practitioners of science were of the opinion that Britain was in danger of being left behind in this promising new field of research. Several prominent figures in this field began calling for a system of magnetic observatories across Britain's imperial possessions. Arguably the loudest of these voices was Edward Sabine (figure 1.1), who had gained extensive experience of making magnetic observations during the Arctic naval expeditions of the 1810s and 1820s. A Royal Artillery officer who was given generous leave from military service to undertake

FIGURE 1.1. Edward Sabine (1788–1883). Portrait by Stephen Pearce, 1855. Image courtesy the Royal Society.

scientific research, Sabine was based at the Royal Military Academy in Woolwich.[6] In addition to an array of fixed observatories worldwide, Sabine also called for an Antarctic naval expedition that would survey Earth's magnetic field in the southern hemisphere and find the as-yet-unknown location of the southern magnetic pole (or poles).[7]

John Cawood, and also Jack Morrell and Arnold Thackray, have claimed that the politically astute Sabine lobbied for this dual project in geomagnetism by putting Humboldt up to writing to the Royal Society, urging Britain to join in the worldwide magnetic campaign and at the same time appealing to British nationalist sentiment by claiming at BAAS meetings that Britain was being left behind in science by its European neighbors.[8] Sabine moved deftly between the Royal Society and BAAS to achieve his aims: when the Royal Society was not initially interested, he took his campaign to BAAS before going back to the Royal Society to seek its authority when applying to the government for funds. In any event, it was John Herschel who during 1838 and 1839 finally secured funding for the Antarctic expedition and magnetic observatories. Fresh from his successful four-year observing expedition at the Cape of Good Hope, Herschel was lionized as a scientific and national hero. He also had class connections at the highest level, which enabled him to lobby for the magnetic project over dinner with Queen Victoria and the prime minister, Lord Melbourne, as well as to negotiate with the aristocratic presidents of both the Royal Society and BAAS.[9] The project that the Melbourne government eventually agreed to fund consisted of an Antarctic expedition under James Clark Ross, running from 1839 to 1842; concurrently with this, magnetic and meteorological observations were to be taken from fixed stations at Greenwich (under Astronomer Royal George Airy), Dublin, Toronto, St Helena, the Cape of Good Hope, and Van Diemen's Land (now Tasmania). Suitable instruments and observational techniques were developed by Humphrey Lloyd, professor at Trinity College Dublin, and the resulting data were collated by Sabine at Woolwich. This combination of an Antarctic expedition and a system of observatories became known as the "Magnetic Crusade."[10]

Herschel, in his 1830 *Preliminary Discourse* quoted previously, seems to have been the first to suggest the general concept of a government-funded observatory to provide long-term data for the use of theoreticians, not just in astronomy but in the physical sciences more generally. Yet nobody could agree on an exact plan for what such an observatory should be doing. The earliest known use of the exact phrase "physical observatory" to describe Herschel's proposal seems to have been made by the Scottish natural philosopher David Brewster, who wrote to William Vernon Harcourt—like Brewster, a leading light in the early years of BAAS—that he had "long thought that one of the greatest scientific desiderata in England is a *physical observatory*, erected and endowed by the government."

Specifically citing Herschel's idea as his inspiration, Brewster suggested that in such an establishment his own experiments in optics could be carried out to a much higher standard than was possible in a private laboratory and that "all the phenomena of magnetism, meteorology and electricity" could be observed as they were in the magnetic observatories then being established across Europe. Harcourt agreed, though such a broadly based concept of a physical observatory made no further progress with BAAS at this time.[11]

Herschel further developed his ideas on physical observatories in October 1835, in a long letter to Francis Beaufort, hydrographer to the Admiralty, written while on his expedition to the Cape of Good Hope. By this time, Herschel had been calling for a more coordinated approach to meteorological observation, both in *Preliminary Discourse* and in the form of an instruction manual for making and recording meteorological observations, originally published in Cape Town.[12] The views expressed in his letter to Beaufort correspond very well with his remarks in *Preliminary Discourse* and are important in that they help us to understand his attitude toward Kew Observatory in the 1840s. Herschel advocated to Beaufort a hierarchical system of observatories worldwide in which the great national observatories such as Greenwich formed a "first class" with which those institutions "of an inferior class" could and should not compete. However, there were many important tasks to be done by these lesser observatories. They should, said Herschel, carry out determinations of constants such as local gravity, mean atmospheric pressure, and sea level (the "absolute height above the level of the Sea of some natural unobliterable mark above or below the station of observatory'). Herschel now also proposed that an important part of these institutions' programs would be to observe, with the most up-to-date instruments and methods available, "magnetic intensity and direction," "meteorology in all its extent," and tides. Thus Herschel's vision of a physical observatory involved routine monitoring of variables such as Earth's magnetic field, as well as the establishment of constants. Herschel had no plans for how such a system of observatories should be put into effect, and he concluded with the reflection, "Perhaps all this is dreaming."[13] We do not know the exact context of this letter to Beaufort, though at the end of the letter he remarks on a ceasefire in the frontier war then taking place in South Africa, suggesting that this vision of a system of observatories was part of Herschel's view of enlightened imperial administration that he developed during his stay at the Cape of Good Hope in the 1830s.[14]

Because of their importance to navigation, geomagnetism and meteorology technically came within the remit of Britain's "first-class" observatory, the Royal Observatory at Greenwich. Edmond Halley (Astronomer Royal 1719–1742) had laid many of the foundations for geomagnetic research, and for some years John Pond, George Airy's predecessor as astronomer royal, had run a magnetic observatory at Greenwich. But the Royal Observatory had never done any magnetic work on a large scale, and by 1835, when Airy succeeded Pond, it had ceased altogether. Airy was a strong supporter of magnetic work, and one of his first priorities on becoming astronomer royal was to set about building a magnetic observatory at Greenwich.[15] Later in the 1830s, he also played a major part in investigating the corrections needed for magnetic compasses on iron ships.[16]

It is clear, however, that from very early on in Airy's time at Greenwich, Airy and Sabine did not get on. This animosity may have arisen because Airy saw Sabine as a rival and a challenge to his authority. It may also stem from the fact that Sabine, unlike Airy and other members of the "Cambridge network" (see the introduction), was no theoretician. Sabine was fundamentally a collector of data who had learned his art through his career in the Royal Artillery and on voyages of exploration. Recruited into the army at the age of fourteen during the Napoleonic Wars, he had not been educated in the regime of reformed Cambridge mathematics in which theory, not empirical data gathering, was seen as the all-important driving force in the physical sciences. Miller has noted the "superior attitude" taken by members of the Cambridge network toward those outside this group. Airy, in particular, was notorious for his insistence on training in higher mathematics as a prerequisite for a leading role in the hierarchy of an observatory.[17] In 1837 Airy refused to support the plan for an Antarctic voyage, apparently out of jealousy toward Sabine's increasing political power.[18] He did agree to take part in the Magnetic Crusade by contributing the Greenwich magnetic observations to the overall magnetic effort. But he was never an enthusiast for the project as a whole, as is emphasized by the sour tone of his letter to a colleague in early 1840: "I have nothing to do with the new magnetic observatories, and know nothing about them. The supreme president over them is Professor Lloyd (Trinity College Dublin) who is certainly willing and I suppose able to tell what they are to be like."[19]

Sabine, lacking the fashionable mathematical education, must have felt himself an outsider in relation to the likes of Airy, Herschel, and

Charles Babbage, who had become such an influential group in the British physical sciences since 1815. Sabine himself made no secret of his preference for a data-driven approach over theory.[20] Moreover, Sabine had been sharply criticized by Babbage in his 1830 polemic, *Reflections on the Decline of Science in England*, for the allegedly dubious accuracy of astronomical measurements obtained during expeditions to measure the figure of Earth. Babbage had also pointed out Sabine as an example of someone holding multiple scientific offices: in this case, acting as both an adviser to the Admiralty and as secretary of the Royal Society while on leave of absence from his army regiment. Much of *Reflections* was an attack on the Royal Society; for Babbage, Sabine's holding of multiple offices was an example of the abuses that he despised in the society.[21] The attitude of members of the Cambridge network toward Sabine might go a long way toward explaining why, throughout the rest of his career, he insisted on being in sole charge of his geomagnetic and other scientific enterprises and would not tolerate interference from Greenwich or anywhere else.

Airy had less regard for meteorology than geomagnetism as a science, because he believed that meteorology lacked a firm theoretical basis.[22] The only meteorological work in the published Greenwich observations before 1840 was a modest set of observations made on the equinoxes and solstices according to the collaborative program that John Herschel had recommended while he was in South Africa.[23] Yet by the time the Kew Observatory building became available in the early 1840s, a fully-functioning magnetic observatory had been established at Greenwich, at government expense. This would have an important bearing on the history of Kew Observatory in the 1840s and its relationship with government.

"I THINK AT KEW": THE ROYAL SOCIETY'S PROPOSAL FOR A PHYSICAL OBSERVATORY, 1840

In June 1840, during the same summer as Forbes's second call for an improved national system of meteorological observations, the Royal Society Council communicated to the government a proposal for something remarkably similar to what was eventually established at Kew: a magnetic and meteorological observatory in the vicinity of London, run by full-time staff and established on a permanent basis. This proposal was in contrast with the system, described earlier, of temporary magnetic observatories set up in various outposts of the British Empire. This episode—which involved a substantial funding application to the highest level of government—has been briefly noted by Marie Boas Hall and also Morrell and Thackray, who have put the failure of the application down to the incompetence of the pre-1847 Royal Society.[24] A further examination of the correspondence reveals some evidence, hitherto unnoticed by historians, that the Royal Society quite possibly had Kew in mind as a location for the proposed observatory. Certainly the similarity of the 1840 proposal to the program of work eventually carried out there means that it is crucial to understanding the history of how the former King's Observatory was transformed into the Kew Observatory of the 1840s and beyond.

The immediate beginning of the narrative can be traced to 4 June 1840, when Forbes, while in London, used the opportunity to launch another of his attacks on the state of British meteorology, particularly the meteorological observations still being carried out at the Royal Society's premises in Somerset House. The Royal Society's Committee of Physics and Meteorology decided to form a subcommittee with a brief to consider and report on this subject. This met four days later and consisted of Forbes, Daniell, and Sabine, plus the meteorologist and electricity expert William Snow Harris. The subcommittee resolved that the observations currently being made at Somerset House were "unavoidably unworthy of the official character which they bear." Its members recommended that this system be replaced with something much more ambitious: the Royal Society Council should now apply to the government "to establish a permanent Meteorological Register in connexion with some National Institution." According to a letter from Forbes to Herschel, at its 4 June meeting the committee had suggested three possible sites where such a "meteorological register" (an improved system of making and recording meteorological observations) might be set up: Greenwich Observatory, Kew Gardens, or Woolwich—the latter presumably meaning the Royal Military Academy where Sabine worked. In a second letter, Forbes reported that on 8 June the subcommittee proposed to locate the new meteorological site at Greenwich on the grounds that such a long-term series of observations was best carried out in a secure, long-established institution—and also, said Forbes, because George Airy had indicated that he would be very willing to take on a "meteorological observer" at Greenwich. Forbes reassured Herschel that the Royal Society's improved system of meteorological observations would not preclude any future proposal to build "physical observatories" of the sort that

Herschel was advocating, which would embrace a much larger program of work.[25]

However, by the next meeting of the Committee of Physics and Meteorology, held on 17 June, the proposal had changed to "a magnetic and meteorological observatory on the same plan as those already established in other parts of the globe . . . in the neighbourhood of London." The council was "recommended to apply to the Government to carry this purpose into immediate effect."[26] This was duly ratified at a council meeting the following day, Thursday 18 June, which was attended by twelve people, among them Edward Sabine. The council requested that the president, Lord Northampton, bring the subject up with the prime minister, Lord Melbourne.[27]

Quite how, between 8 and 17 June, the plan was transformed from an improved version of the Royal Society's meteorological record into a full-blown observatory doing magnetic work as well as meteorology, is not clear. However, some clues can be found in a brief exchange of letters between George Airy—who was not a member of the Royal Society Council or the Committee of Physics and Meteorology—and his old friend Richard Sheepshanks, a noted astronomer and a member of the Royal Observatory's Board of Visitors. Sheepshanks reported that while walking home from the Athenaeum club on the evening of Monday 15 June, he had been informed by Ordnance Survey director Thomas Colby that "there was some talk of the want of magnetic observatories at Greenwich & that there was & would be a considerable difficulty as to the regulation of the magnetic observatories recently established" (those of the Magnetic Crusade). In addition to this slight against Airy's magnetic establishment at Greenwich, Sheepshanks's informer also intimated that there was the possibility of a move to appoint a "magnetic chief"—he named Sabine as a possible candidate—to run the magnetic observatories independently of Greenwich.[28] Sheepshanks's letter has the tone of a friendly warning to Airy, who seems to have been kept in the dark about the whole move. From this, it seems possible that it was Sabine who turned the idea for an improved meteorological register into a magnetic as well as meteorological observatory. Sabine had attended all the various committee and council meetings between 4 and 18 June. Moreover, with his long experience in magnetic survey work, he was an obvious choice for the post of "magnetic chief."

The application took the form, to begin with, of a deputation consisting of Lord Northampton, Edward Sabine, Royal Society treasurer John

Lubbock, and Samuel Hunter Christie, secretary of the Royal Society and a professor at the Royal Military Academy in Woolwich. These four visited Downing Street on 20 June—a Saturday, when Melbourne presumably had more time for such visitors. John Herschel did not attend this meeting or any of the committee or council meetings in June 1840. He was not a council member by this time; in addition, just two months earlier he had moved into his secluded country residence in Kent and so he may well have been preoccupied with settling in.

The day after the meeting, Lubbock reported that the prime minister had received the visitors well and that although no decision could be made there and then, Lubbock had "no doubt that sooner or later it [funding] will be granted." More importantly, Lubbock's letter clearly shows that the Royal Society was seeking funding for something very different from the temporary observatories of the Magnetic Crusade, which were to have an initial lifetime of three years. The proposed new observatory would be "a permanent magnetic and meteor. observatory."[29] According to a later letter from Lubbock, the observatory would need a director plus three assistants, resulting in a total annual salary cost of £2,000. The cost of printing the observations, together with various other expenses such as repairs, increased the cost estimate to a minimum of £3,000 per year. Lubbock emphatically stated that in addition to magnetic work, the observatory would also carry out "meteorological observations similar to those now made at the Royal Society but on a more extended system"; in addition, he noted that "it may be desirable to devise also observations of the electrical state of the air & others which the Royal Society did not furnish."[30]

Sabine, in a letter to Herschel, claimed to have been "perfectly ignorant of what had passed at the Council" on 18 June and that Lubbock and the others had put him on the spot, forcing him to say to Melbourne that the Royal Society wanted a permanent observatory. Sabine's name clearly appears on the list of those who attended the council meeting, so he could not have been as "perfectly ignorant" of what had happened as he claimed.[31] As Herschel was not a council member, he would not have had access to the minutes of its meetings, and so it would appear that Sabine was not telling the truth. Sabine might also have been lying in his claim to have been put on the spot by the others at the meeting with the prime minister: no surviving letters or minutes before 20 June suggest that the new observatory would be anything other than permanent, and on no other occasion did Sabine express misgivings as to the new observatory being a permanent establishment. Later in 1840 Airy, admittedly no friend of Sabine's, claimed that the proposal had been presented to the government as it had due to Sabine's reporting the view of Berlin astronomer Johann Franz Encke that the magnetic and other work would be too much for Greenwich and so a separate observatory would be required.[32] If true, this suggests that Sabine might have been using the opinion of a well-known continental astronomer to strengthen the case for a separate observatory. These factors, plus Sheepshanks's letter to Airy, as well as Sabine's track record of a poor working relationship with Airy, suggest that there is at least circumstantial evidence of Sabine colluding in the idea of a permanent observatory independent of Greenwich and that he might well have been central to the project.

Exactly how George Airy came to hear about the deputation to the government is not known. Certainly he was tipped off by Richard Sheepshanks on 17 June about the Royal Society's plans, but by the time he wrote to Lord Northampton on 28 June he had more up-to-date knowledge: "I have just heard a very vague report that a recommendation has been addressed to the Government by the Council or by a Committee of the Royal Society, to the effect that a magnetic observatory should be erected or fitted up by the Government, I think at Kew." Airy went on to describe his "excellent Magnetic Observatory" at Greenwich, which had been built "at considerable expense to the Government." Asking Northampton for further information on the proposed new observatory, he emphasized the importance of saving the government expenses that were "absolutely unnecessary" and that "the machinery of a new establishment should be dispensed with when that of an old one can be made available."[33]

Airy's 28 June letter does not prove that the former King's Observatory was to be used as a site for the proposed new magnetic and meteorological establishment. Indeed, just before the deputation Sabine claimed to Herschel that "no one . . . seems to have any distinct idea of where such an establishment can best be placed."[34] Yet according to Forbes's 4 June letter, Kew Gardens had been one of three possible sites for the original suggestion of an improved meteorological "register."[35] Moreover, it is interesting that Lubbock's cost estimate details only the annual running costs and makes no mention of a suitable building or instruments. The latter might have been paid for out of private funds or the Royal Society's own coffers, but erecting a new, permanent building from scratch would have required a substantial capital investment beyond private or Royal

Society means. If the Royal Society had in mind a ready-made building that was available free of charge, it is difficult to think of any building at Kew other than the King's Observatory. In fact, there is evidence that several leading men of science had been alerted to the situation of the disused observatory in the weeks following the death in 1839 of Stephen Rigaud, who had assisted with the running of the observatory. Shortly after Rigaud died, the astronomer William Rutter Dawes had written to Herschel and Airy about applying for Rigaud's vacant position at Kew.[36]

In a second letter to the Royal Society, Airy outlined an alternative plan: an extended magnetic and meteorological observatory at Greenwich, under his own direction, which would obviate the need for a new, separate establishment. Airy offered to use his existing magnetic building and to carry out the same program of work with fewer extra staff than the Royal Society's proposal, sharing some personnel with the main astronomical observatory. His letter was read out at a meeting of the Committee of Physics and Meteorology on 9 July and at a meeting of the full council the same day. The minutes of the council meeting quote a total extra staff cost of £550 per annum. Even if we add the costs of extra instruments and printing, the total cost was obviously far less than the £3,000 per year quoted in Lubbock's plan. Again, Airy made much of the need to save public money: "As the Government have been led to expend a considerable sum on this building for this purpose, . . . I do think that it would be most unfair towards the Government, and most injurious to the cause of science in future negociations [sic] with the Government, to set aside the consideration of this investment in judging what is best to do at present." Airy concluded his letter by asking Lubbock to "assure the Committee that, if they determine on not accepting my offer, I shall fully understand that the inconveniences attached to it do in their estimation exceed the conveniences."[37] Airy may have sincerely thought that the committee might find his offer of an extended magnetic observatory inconvenient, perhaps for practical reasons, Greenwich being some distance from central London. But these words might have been a polite way of sending a different signal: if the committee were to reject Airy's offer, Airy would take it that its members thought that he could not do as good a job as the Royal Society. The latter interpretation is especially plausible given Sheepshanks's warning about "talk of the want of magnetic observatories at Greenwich," which would surely have perturbed Airy. Both of Airy's letters have a clear tone of anxiety about the very idea of a separate observatory. Clearly he wanted to keep the permanent magnetic and meteorological observations under his own control and saw any separate observatory as a rival.[38]

Faced with Airy's offer to do the job more cheaply and just as efficiently, the Royal Society had no option but to back down. At the council's request, Northampton wrote to Melbourne on or before 20 July, retracting the application made one month before, claiming that the request for a separate observatory had been due to concerns about Airy's lack of resources to do the extra work himself but he was now pleased "to find that we were mistaken as there can be no doubt of his entire fitness for the most satisfactory performance of such additional duties."[39] The government agreed to fund an extended magnetic and meteorological observatory at Greenwich, where it became known as the magnetic and meteorological department.[40]

Melbourne was no doubt relieved at not having to fund the new observatory, particularly as by 1840 he was leading a minority government. Only a year earlier his government had reluctantly agreed to support the extremely expensive Antarctic expedition and the accompanying magnetic observatories, so it must have been difficult for him to justify yet more spending on costly scientific projects—in this case, moreover, a permanent observatory, not a one-off series of temporary ones. Indeed, the political situation might well explain why the Royal Society acted with such haste in applying for funding in June 1840: if it did not move quickly, the government could fall and be replaced by a Tory administration under Sir Robert Peel, who had a reputation for being keen to reduce public spending.

Herschel might well have agreed with the prime minister about yet another substantial application for funding so soon after the Magnetic Crusade: "It would not only *seem* but *be* importunate to press, just at present for further grants in this direction." Herschel strongly supported, in principle, the idea of a permanent magnetic and meteorological observatory: he thought that such an institution would "do honour to the country & confer great benefits on science." According to Herschel, even more important was what he termed an "experimental Institute or College" that would do more general standardization work: "an institution destined for the systematic determination . . . of all the invariable elementary data of physical theories which admits of such determination such as atomic weights—specific heats—pyrometric changes—electric—thermotic &c constants—such that is to say as are *not* of a local or temporary nature." Yet most tellingly for the future of Kew Observatory,

Herschel thought that "the proper *locale* of a physical observatory should be on the Sea Coast—1st for observation of the tides—2d [*sic*] as a centre of departure of a general coastland-line—to be ultimately referred to the mean-sea level at that spot as a probably invariable standard."[41] So it would appear that Herschel had good scientific reasons for not supporting Kew or anywhere else near London as a good location for a physical observatory: this was obviously the wrong location for a coastal observatory. Moreover, the vision expressed here of an observatory at a coastal location, measuring physical constants as well as making magnetic and meteorological observations, is entirely consistent with Herschel's earlier ideas for physical observatories in *Preliminary Discourse* and his October 1835 letter to Beaufort.

Thus the Royal Society's failed application for a new government-funded observatory was likely a carefully planned maneuver by Sabine that was foiled only by Airy and his intelligence network. A few months later, however, there would be a new possibility of an observatory at Kew, and this time Airy would be powerless to do anything about it.

KEW OBSERVATORY AND THE ROYAL SOCIETY, 1841–1842

On 5 February 1841, Sabine wrote to Herschel with news from Francis Beaufort: a government official had told Beaufort that the former Kew Observatory building was in such excellent condition that it would not be pulled down, as had been intended, and had asked him if he could think of any use for it. According to Sabine, Beaufort had suggested a magnetic observatory, to which the official replied, "Well, so, so, & you will have it, most likely." Sabine went on to say, "For altho' the arrangement relative to Greenwich seems to have forestalled the use that could so well have been made of it as a Magnetc. & Meteor. Observatory, it seems a very suitable place for your ulterior project of a Physical Observatory."[42] This adds further weight to the idea that Sabine, at least, had had the Kew building in mind in 1840. Beaufort had certainly been aware for some time that the building was unused: as early as 1839 he had reported to Herschel that it was to be pulled down.[43]

Nothing further then happened until 24 June. It might have been Sabine, who attended the meetings of the Committee of Physics and Meteorology and full council that day, who informed the Royal Society that Kew Observatory was being made available by the government, apparently free of charge. In any event, the Committee of Physics and Meteorology passed a resolution in favor of acquiring it. The council duly adopted the resolution, and once again the president was requested to make an application to the government—though not for funding this time, only for possession of the building.[44] But Northampton does not appear to have done this, and instead there was another long delay. Nearly five months later, the council asked the committee to report back as to for "what specific scientific purposes it would be desirable to appropriate the building formerly occupied by the Observatory at Kew" and to suggest "what would be the probable annual expense of applying it to such purposes."[45] The committee duly appointed a subcommittee consisting of Herschel, Sabine, and Charles Wheatstone (professor of experimental philosophy at King's College, London, since 1834) to draw up the report for the council. These three met on 18 December, though their resulting report was not read to the council until 10 February 1842. The report gave a mixed verdict on the observatory. To begin with, the subcommittee thought that Kew was not suitable for "any regular and systematic course of physical observations" by the society due to its "peculiar restrictions as to access and inhabitancy and other circumstances." The report did, though, recommend several other uses for the building, such as a depository for Royal Society instruments and a place for comparison of instruments such as pendulums. The estimated costs were a salary of about £27 per annum for a caretaker and a mere £5 per annum for maintenance—a far cry from the £3,000 annual cost for the 1840 observatory and even the £550 for Airy's extended magnetic and meteorological establishment at Greenwich.[46]

That the report dismissed Kew as unsuitable for regular, systematic observations might at first seem surprising, given that the building would be used for precisely that purpose later in the 1840s. There were indeed some genuine "restrictions to access and inhabitancy," such as the building's remoteness from central London before the arrival of the railway and the fact that an existing caretaker already occupied the basement, but these were not insurmountable. We do know, however, that the report was drafted by Herschel,[47] who, as already discussed, would not have considered Kew a good site for a physical observatory. Probably no one at this time had greater authority in the physical sciences than Herschel, and in a subcommittee of just three people Sabine and Wheatstone might have had no choice but to defer to his wishes. But the main reason why the Royal Society decided not to use Kew as a magnetic and meteorological observatory—and, indeed, why the council did not immediately go ahead with the proposal to acquire the building—might well have been financial. Given that the council had specifically asked about the "annual expense" and that the total cost of the watered-down proposal amounted to little more than £30, it is likely that a full-scale observatory, complete with staff and instruments, would have been too large an annual charge on the Royal Society's funds.

Herschel himself might well have shared the general consensus about costs: just before the December 1841 subcommittee meeting, he confessed to Sabine that he thought Kew Observatory "likely to cause some degree of embarrassment" to the Royal Society.[48] If Kew were to cost a significant amount of money to run, the Royal Society would once again have had to apply for a hefty government grant little more than eighteen months after its retreat in 1840. Also, Herschel believed that large-scale physical observatories of the sort envisaged in *Preliminary Discourse* should be run by the government, not scientific societies: much later, he expressed the belief that to take responsibility for an observatory or any other permanent institution would "deprecate" the Royal Society.[49] To make matters worse, by the end of 1841 the political climate had changed: Melbourne's Whig government had finally fallen and had been succeeded by the Tories under Peel. Herschel, for one, considered the outlook for science under Peel's government "exceedingly ill-omened" and bemoaned "the *good old Tory feeling* of hatred and contempt for Science and its followers."[50]

It is also possible that Sabine, who had ruthlessly used both the Royal Society and the BAAS in his lobbying campaign for the Magnetic Crusade, agreed to the watering down of the proposal in order to steer the Royal Society toward rejecting the government's offer, with the ulterior motive of making the observatory available to BAAS, which he might well have thought would be more receptive toward it. We do not have any documentary evidence as to Sabine's and Wheatstone's motives, however. All we know is that at the council meeting exactly one month later, with no reasons recorded other than consideration of the subcommittee's report, it was decided that "it does not appear to the Council to be expedient for the Society to occupy the Observatory at Kew." The council requested the treasurer, Lubbock, to communicate this decision to the government.[51]

THE BRITISH ASSOCIATION:
FOUNDING THE "ESTABLISHMENT" AT KEW, 1842–1843

On 28 March 1842, just eighteen days after the Kew building was finally rejected by the Royal Society, the possibility of acquiring it for BAAS was formally raised at a BAAS Council meeting. Roderick Murchison, general secretary of BAAS, noted that the Royal Society had declined the offer of the building, "and that if an application should appear desirable on the part of the British Association, it was necessary that it should be made without delay." Sabine and Wheatstone were both present at this meeting; indeed, Sabine attended most of the BAAS Council meetings over the next several years. At the 28 March meeting Wheatstone, who had been on the Royal Society subcommittee that had rejected Kew as a site for systematic observations, now read a statement "of several important objects in the Physical Sciences," which the Kew building would offer to BAAS members "in the prosecution of experimental inquiries."[52]

Wheatstone appears to have drawn up this document, and it is apparent from it that the proposed program of "experimental inquiries" was very different from the Royal Society's watered-down proposal. It stated unequivocally, "It is proposed to establish, in connexion with the British Association, a Physical Observatory" in the Kew building. The objectives of this physical observatory fell under seven broad headings: a repository "and place for occasional observation and comparison" of newly invented meteorological instruments; the construction and trial of new self-recording meteorological instruments; a repository of standard instruments with which people could compare their own instruments; a place where magnetic instruments currently used "in the various magnetic observatories" could be kept to enable people to learn how to use them; the setting up of apparatus for research into atmospheric electricity; a room for experimental work on optical astronomical instruments (an echo here of David Brewster's 1832 call for a place to carry out his optical experiments); and a collection of measuring instruments, "for the purpose of obtaining accurate quantitative results."[53] The 28 March council meeting quickly approved the proposal. On 16 May a formal application was sent to the prime minister, and just ten days later the government sent an official letter to BAAS to the effect that the Queen had given her permission for the association to take possession of the building.[54]

The contrast between the response of the Royal Society and that of BAAS to the Kew offer is dramatic: whereas the Royal Society's discussions took nine months, BAAS made the decision at the same meeting at which the availability of the building was announced and took possession of the observatory just over two months later. This further strengthens the possibility that Sabine had given up on the Royal Society as a probable lost cause long before the formal rejection on 10 March—and even that he had prepared the ground with colleagues on the BAAS Council well before the meeting on 28 March. But without a record of what was actually said at the meetings we cannot know for sure. Certainly there is no record in the BAAS Council minutes of anything being discussed about Kew Observatory in the months prior to 28 March 1842.

Decisions by the BAAS Council to take on new projects normally had to be sanctioned at one of the association's annual meetings. On this occasion, however, the council resolved to take possession of Kew well before the 1842 annual meeting, which was held in Manchester in late June of that year. This may have been to avoid doubts creeping in if proceedings were delayed, as had happened at the Royal Society, or to preempt any dissent at the annual meeting. The acquisition of Kew was duly announced to the membership and wider public at the annual meeting on 22 June, and it was approved with no recorded dissent. Even more telling was the vote of £200 to "be placed at the disposal of the Council for upholding the establishment in the Kew Observatory."[55] Not only was this a very different sum of money from the approximately £32 a year suggested by the Royal Society, the phrase "upholding the establishment" is suggestive of a permanent, or at least long-term, institution. And indeed, BAAS voted for similar sums of money for Kew over the next few years: £200 in 1843 and £150 in 1844 and 1845.[56] Most importantly, this annual vote was not a government grant but was from BAAS's own limited resources, which further underlines the commitment given to the project by Sabine, Wheatstone, and the others on the BAAS Council. In the early to mid-1840s BAAS made no proposals to apply for government funding for Kew, nor is there evidence of any such proposals being considered at this stage. It was purely a privately funded project.

John Herschel seems to have played no part in the British Association's acquisition of Kew Observatory. In fact, BAAS made more than one appeal to his authority during this time, not only to seek his advice but also, one feels, to obtain the backing of someone who was seen as the leading figure in the physical sciences in this period. This was certainly the tone of a letter from Murchison to Herschel, imploring him to attend the 1842 annual meeting: "On *this occasion* your presence would be doubly useful in helping us to give birth to the child which you have so large a share in creating—the Kew Observatory of Physical Science."[57] Herschel did attend this meeting, but he did not become personally involved in any BAAS committees on Kew at this stage. In reply to a letter from Wheatstone enclosing a draft of his prospectus for Kew Observatory, he expressed no particular disagreements with the project and thought that the observatory would be useful for experimental work, but he was rather cool toward the whole idea. It seemed to Herschel "not very clear" that the British Association's plan for Kew as a physical observatory would work. He doubted whether BAAS had adequate funds to support a physical observatory that did long-term, systematic observations for the production of data useful in theoretical work.[58] Yet he very likely had the same doubts about the Royal Society taking on such a heavy annual budget commitment. Moreover, as in 1840 he now also questioned whether "*the locality is fitted*" (Herschel's emphasis) for such purposes.

But perhaps the main reason why Herschel did not want to become too closely associated with Kew was that by now he was reluctant to become heavily involved in the management of large scientific projects generally. Always preferring to do research in a private capacity without any obligation to larger organizations or committing himself to regular, time-consuming work, Herschel was now fifty years old and anxious to settle down to the mammoth task of writing up the results of his astronomical observations at the Cape of Good Hope while he still had time and physical energy left. In the same June 1842 letter to Wheatstone, he expressed the wish to confine himself to "general advocacy" of scientific projects except for those that he felt particularly passionate about, "now . . . that I can calculate on but very few years more of scientific efficiency."[59] BAAS was left to commence its program of observational work at Kew without Herschel's active involvement.

FRANCIS RONALDS, METEOROLOGY, AND ATMOSPHERIC ELECTRICITY AT KEW

BAAS lost little time in preparing the newly acquired building for work. In July 1842, the BAAS Council appointed a committee "to superintend for the present the arrangements at the Kew Observatory." This consisted of Wheatstone, the two general secretaries of the association (Murchison and Sabine), and the treasurer.[60] In charge of the day-to-day work at the observatory for its first ten years under BAAS was Francis Ronalds, a renowned inventor who had developed an early form of electric telegraph in 1816–more than twenty years before Wheatstone, in partnership with James Fothergill Cooke, patented the commercially successful five-needle telegraph. Little is recorded as to how Ronalds, now in his fifties, was appointed to direct the association's newly acquired institution, though by the early 1840s he was respected in scientific circles and had known Wheatstone for many years. According to an autobiographical letter dated 1860, Ronalds was offered the post by BAAS: keen to return to his interests in electricity and meteorology, he "accepted the honorary direction of the hardly more than projected Meteorological Kew Observatory under the auspices of the British Association." Ronalds was from a comfortably well-off family of cheese merchants and had funded his own research since he was a young man.[61] It is an indication of the association's limited budget that, in contrast to all his successors at Kew, Ronalds drew no salary for his post. That Ronalds did not require a salary must have made his appointment attractive to the committee, with its very small initial budget of £200. Ronalds was the first director of Kew Observatory to be known as its "superintendent," in contrast to the title "King's Observer" used by both his predecessors.

In January 1843, the BAAS Council announced that it had employed an assistant, John Galloway, to take care of the observatory, to help BAAS members doing research there "and to obey to the best of his ability whatever instructions he may receive from time to time." Galloway was initially paid an annual salary of £27 7s. 6d. and was arranged living accommodation in the building. From the beginning he was much more than a caretaker. From 1 November 1842, he used instruments purchased by BAAS to keep a systematic record of meteorological observations.[62] We have no formal record of Galloway's background or what, if any, scientific training he possessed, but in his 1844 report to BAAS Ronalds describes a new anemometer, attached to which is a "sentry box," "the invention of Sergeant Galloway, who made nearly the whole instrument."[63] Given that Edward Sabine employed soldiers to perform the day-to-day instrument readings in the colonial magnetic observatories, it is quite possible that Galloway was a soldier or ex-soldier recruited by Sabine from among his subordinates in the Royal Artillery at Woolwich. This possibility is greatly strengthened by Ronalds's earlier remark to Wheatstone that "I suppose that the Artillery Sergeant could do some of the heavier work which might be wanted."[64] The employment of Galloway also makes it more likely that Sabine was central to the whole Kew project.

According to Ronalds's 1844 report, the meteorological record begun by Galloway measured temperature, pressure, humidity, rainfall, wind speed, and wind direction. Observations were made at least twice a day, "almost exclusively by Mr. Galloway." It is notable that from the beginning, high-quality instruments were used. When they could not be afforded, they were borrowed, as with a "mountain" (portable) barometer "lent by Colonel Sabine until we can afford the expense of a standard instrument." Even more important, from the beginning of his reports Ronalds showed a critical attitude toward both his instruments and his observations. Where possible, instruments of different types were used at the same time and results compared. Those whose accuracy was found to be wanting were dropped. With regard to the observations, Ronalds praised Galloway's efforts but reflected that "had our habits and qual-ifications been always adequate to the attainment of extreme accuracy, our instruments and other means would have been far from being so."[65] This comment suggests that, according to Ronalds, the instruments were only as good as the less-than-perfect observers who used them. It is clear that Ronalds was trying to do meteorology to the highest possible stan-dard of accuracy, perhaps higher than had hitherto been achieved any-where else.

The second item in the 1842 prospectus suggested that Kew should become a center for building and testing self-recording meteorological instruments. Automatic meteorological instruments were nothing new by 1842: self-recording barographs and thermographs, automatically re-cording observations on rolls of paper, had been in occasional use since the late seventeenth century.[66] However, automation of meteorological (and astronomical) observations was coming into vogue by the 1840s, as the new technologies of telegraphy and photography greatly extended the possibilities in this field. In 1839, the year in which Louis Daguerre and William Henry Fox Talbot first announced their photographic pro-cesses, came the demonstration of a barograph that recorded a trace onto photographic paper. In the same year, Scottish astronomer John Pringle Nichol called for photographic registration to be used more widely in meteorological observations.[67] In 1844 the BAAS Council authorized the expenditure of £30 for the purchase of a top-quality self-recording barometer by Karl Kreil of Prague, and a further £25 was spent on transferring it to Kew[68]—another sign of the association's commitment to using the very best instruments at the observatory. Even more im-pressive was Wheatstone's "Electro-magnetic Meteorological Register," which automatically recorded 1,008 observations per week. It contained instruments for recording temperature, pressure, and humidity, each of which was activated in turn when a wire connected to the top of the mer-cury in the instrument sent a signal to two type wheels, which printed the instrument reading in figures. Yet although Wheatstone's six-foot-high device was pioneering and must have been a spectacular example of instrumental innovation and prestige at Kew,[69] it did not replace traditional meteorological observations and instruments. Rather, it was experimental in nature. Although experimentation was clearly on the agenda, Kew was becoming at least as much a central meteorological observatory as it was an experimental station.

From the summer of 1843, Ronalds and Galloway also began to make observations of atmospheric electricity, which had been stated as a clear objective in both the 1840 proposal and the 1842 prospectus. These electrical observations were recorded along with the traditional meteorological readings and take up about half of the columns in the meteorological record as reproduced in the 1844 BAAS report. The observations were made in the observatory dome; according to Ronalds's 1844 report, the instruments used to make the measurements were attached to the base of a conductor, a sixteen-foot-long tube of copper placed vertically so that it protruded twelve feet above the dome's outer surface. Observations were made four times a day of the intensity of electric charge and whether this was positive or negative. In addition, a maximum and minimum charge was noted, based on hourly observations between noon and 10 p.m., and an attempt was also made to relate the electric charge to the type of weather.[70] These electrical observations must have made for a demanding routine, for in addition to the meteorological readings Galloway had to read the electrical instruments "every day from half an hour before sunrise until night." In return for this, his salary was increased to one guinea per week, or almost double his original remuneration,[71] which demonstrates how seriously BAAS, with its limited budget, was taking this work.

On a first reading, Wheatstone's 1842 prospectus—unlike the 1840 proposal—makes no provision for magnetic observations at Kew. It merely mentions that the observatory could be used as a place for storing magnetic instruments and training observers in their use. But the prospectus clearly did not preclude systematic observational work, for even the electrical observations are described therein only as "experiments on atmospheric electricity."[72] In any case, the electrical observations had an important connection with geomagnetism. Apparatus for such observations also formed part of the equipment of the Magnetic Crusade observatories. According to the 1843 BAAS report, the committee in charge of Kew Observatory noted that atmospheric electricity had been given priority "on account of its importance in connexion with the system of simultaneous magnetic and meteorological observations now making on various points of the earth's surface, in the recommendation of which the Association has taken so prominent a part."[73] Thus the Kew observations of atmospheric electricity were directly related to the Magnetic Crusade. Moreover, as early as November 1842, in a list of meteorological instruments he said were needed at Kew, Ronalds had asked for "Dipping & Variation needles"[74] and his 1844 report includes an as-yet empty column in his meteorological register "intended for the deviations of the electro-magnetic needle,"[75] strongly suggesting that at least basic magnetic observations were being planned for the near future, perhaps when funding for instruments was forthcoming. Sabine may even have applied for a grant from BAAS for magnetic work at Kew in 1842 (something not mentioned in the prospectus), for a private letter from Wheatstone mentions a "proposition for the grant for the magnetic instruments."[76]

Sabine and Herschel succeeded in persuading Robert Peel's government to renew funding for the Magnetic Crusade in 1842 and again in 1845. The money for this project only finally ran out in 1848.[77] Yet despite Sabine's central role in starting observational work at Kew Observatory under BAAS, and notwithstanding the clear link between the Kew electrical observations and the larger magnetic project, Kew was never really part of the Magnetic Crusade in the way that the magnetic observatories at Greenwich, Dublin, and the colonies were. No government funding for the Magnetic Crusade ever went to Kew in the 1840s. The resulting data were collated and analyzed by Sabine and other soldiers at Woolwich, not at Kew. The electrical observations at Kew, paid for out of BAAS's limited funds, seem to have been no more than supplementary to the work of the colonial observatories. Nevertheless, the work in atmospheric electricity, as well as the evidence for planned future magnetic observations, together strongly suggests that the 1842 prospectus did not prevent Sabine from trying hard to slip his beloved magnetic observations into Kew by the back door.

CONCLUSION

In his 1842 prospectus for Kew Observatory, Wheatstone claimed that BAAS wanted to turn the former King's Observatory into a "physical observatory" dedicated to meteorological observation, work in atmospheric electricity, and experiments with new types of self-recording instruments. Yet only up to a point was it a physical observatory of the kind proposed by John Herschel. While aspects of it—meteorology, atmospheric electricity, and experimental work—were certainly Herschelian, it was clearly not the central observatory that Herschel had in mind, which would also have incorporated fundamental work such as tides and sea levels and, in Herschel's view, would have been a government institution, not privately run by BAAS (or the Royal Society). This goes a long way toward explaining why Herschel was equally lukewarm about Kew with both the Royal Society and BAAS: the building at Kew was in entirely the wrong location for his idea of a physical observatory, and both organizations, he felt, were incapable of providing adequate financial support for such an institution.

There is good evidence that the prime mover behind the whole Kew project, at every stage from June 1840 onward, was not Herschel but Sabine. As we have seen, Sabine had a motive: due to his dislike of Airy and other members of the Cambridge network, he wanted to assume control of the magnetic and meteorological observations. The hand of Sabine is visible time and again throughout the story. That Sabine was behind the 1840 proposal is strongly suggested by the maneuverings behind the scenes in the summer of that year. It was Sabine who, early in 1841, first let the Royal Society know of the availability of the Kew Observatory building and who then, seeing the society's lack of enthusiasm, was one of those who took the project to BAAS, perhaps deliberately steering it toward the latter organization. At any rate, Kew in the mid-1840s was a permanent "establishment" (BAAS's own word) and was essentially a meteorological observatory, having as a central part of its program observations of atmospheric electricity tied to the Magnetic Crusade—albeit without the observatory being fully a part of the larger magnetic project. In other words, it pursued an agenda consistent with the 1840 proposal for a magnetic, meteorological, and electrical observatory independent of Greenwich, as far as was possible in the absence of government funding. Had he been able to secure the funds, Sabine would no doubt have set about establishing Kew as the main British observatory in the Magnetic Crusade. Indeed, some years later Sabine confessed privately to the meteorologist and astronomer William Radcliff Birt that the government, by means of "observations at Greenwich," had "undertaken to do, and in the most efficient manner what we wished to have done at Kew but what we have never been able to accomplish except in a degree very inferior to our wishes."[78] Sabine, whom one historian has described as "the artful dodger of the British scientific establishment,"[79] had manipulated both the Royal Society and BAAS toward his own agenda. While it is easy to see the establishment of an alleged "physical observatory" at Kew as a straightforward realization of Herschel's dream, in reality the story is a more complex one of personal agendas and the realities of funding large scientific projects in the 1840s.

Survival and Expansion

KEW OBSERVATORY, THE GOVERNMENT GRANT, AND STANDARDIZATION, 1845–1859

The Committee are glad to have an opportunity to testify to the increasing utility of the operations at the Kew Observatory, in the very laborious verifications, by Mr. Welsh, of the twenty sets of Meteorological Instruments intended by the East India Company for proposed meteorological observations in India.

REPORT OF THE KEW COMMITTEE,
READ AT BAAS COUNCIL MEETING,
31 JANUARY 1852

KEW OBSERVATORY IN THE EARLY 1840S WAS SUPPORTED ENTIRELY BY BAAS's limited funds. In 1845, the observatory had an annual budget of (at most) £200 and a restricted (albeit very definite) program of work. Yet by 1859, it was an internationally recognized institution whose budget had almost quadrupled. By then, moreover, part of its work was self-financing and on a commercial basis. The observatory was also headed by a full-time, paid superintendent with scientific qualifications. This chapter shows how this transformation happened. It came about partly due to an increase in the grant it received from BAAS, plus the introduction of some limited government funding. However, the commercial astuteness of two key BAAS members, plus the technical ingenuity of several individuals working at Kew, were also key to this huge and rapid expansion of the observatory.

The first section discusses how Kew Observatory survived several threats of closure in the 1845–1852 period. This was partly due to Edward Sabine's political astuteness: as he did during the moves to acquire Kew for BAAS in the early 1840s, Sabine presented an official mission for the observatory that was politically acceptable while all the time pursuing his private agenda for a magnetic and meteorological observatory independent of Greenwich. The advent of the Royal Society government grant in 1850 helped the observatory to survive—though not in the sense of the government supporting it with an annual grant. However, the BAAS Council may have delayed closing the observatory in anticipation of government support that had not yet been publicly announced. The second section describes another important factor in saving Kew from closure: the introduction of instrument testing on a commercial basis. This brought the observatory a substantial extra income and, most importantly of all, a worldwide reputation for making and testing standardized meteorological and magnetic instruments. The standardization program made Kew indispensable to a range of organizations far beyond the Royal Society and BAAS—by the end of the 1850s, there was no talk of closing it down.

Also during the 1845–1859 period, the range of observatory sciences practiced at Kew notably expanded from the limited program of work that was in progress during the mid-1840s. The last section of this chapter describes how Sabine finally established his cherished magnetic work at Kew so that by the late 1850s Kew was a world center for geomagnetic observations. This diversified range of geomagnetic and meteorological observations itself increased the observatory's prestige—as was demonstrated, for example, by a government-funded display of the Kew instruments at the 1855 Universal Exposition in Paris. The growing program of work required extra staff; as a result, by the end of the 1850s Kew had several staff, some of them with scientific training. In its development between 1845 and 1859, Kew mirrored the trend, identified by David Aubin, of diversification in the range of sciences practiced in observatories during the mid-nineteenth century.[1] Indeed, it will become clear that Kew did not merely follow this trend: along with its work in standardization, Kew itself became a prominent landmark in the mid-nineteenth-century observatory sciences.

SURVIVAL OF KEW OBSERVATORY, 1845–1852

At each of its annual meetings from 1842 to 1845, BAAS consistently voted grants of between £150 and £200 for the running of Kew Observatory. But at the 1845 meeting the BAAS General Committee passed the following motion: "That it be referred to the Council to take into consideration previous to the next Meeting the expediency of discontinuing the Kew Observatory."[2] As is typical of the BAAS annual reports, no reason was given as to why this recommendation was passed. However, a report in the *Times* the next day noted that "a long discussion took place upon the propriety of the discontinuance of the Kew observatory, on the ground that it had been ascertained that the observations there being carried on formed part of the subjects of observation at Greenwich. It appeared that the expenses of this establishment were 150*l*. per year, which the reduced funds of the association would not allow."[3] The article was anonymously authored and it appeared in a national newspaper that had a long history of being critical of BAAS.[4] It therefore speaks more freely than any report from within the association, and even though we have no idea as to whether the author was present at the meeting or working from secondhand reports, it offers more insight into what might have happened. A similarly anonymous report in the *Athenaeum* notes that the council was asked "to consider whether the Electrical Experiments at the Kew Observatory should not be discontinued." The *Athenaeum* claimed that the decision had nothing to do with money but again said that it had been taken because "similar observations were now being made at the Observatory at Greenwich, under the superintendence of Prof. Airy."[5]

It is certainly true that by the 1840s BAAS, still reliant on income from subscriptions and voluntary donations, was finding it increasingly difficult to meet its financial commitments. It had always been the association's policy to fund one-off projects in preference to permanent institutions, and in the competition for grant money Kew had many rivals from across the whole range of sciences. But regardless of any financial motive, both these independent sources cite the same primary reason for considering closure: that Kew was duplicating work being done at Greenwich. We have no documentary evidence as to who originated the motion to close down Kew, but it is reasonable to speculate that it might have been another move by George Airy to put an end once and for all to this source of competition from the other side of London. The 1845 annual meeting was held at Cambridge, where Airy had found his intellectual feet in the world of reformed Cambridge mathematics in the 1810s and 1820s.[6] Many of his old friends from the university were present at this meeting and played leading roles—notably John Herschel, who served as president. Airy had been observing atmospheric electricity at Greenwich since at least March 1842,[7] so he could quite reasonably have claimed that the work at Kew was duplicating that at Greenwich. It is easy to imagine Airy being in a strong position at the 1845 meeting to argue against the continuance of Kew Observatory.

There is no evidence that the June 1845 motion raised any particular alarm bells with Sabine or the other advocates of the Kew project. The BAAS Council only came around to the issue at its January 1846 meeting, when it appointed a committee "to collect information on the scientific purposes which the Kew Observatory has served, and on its general usefulness to science and to the Association." Herschel was to chair the committee; its other members were Airy, Thomas Graham (professor of chemistry at University College London), George Peacock (Airy's former Cambridge tutor, now dean of Ely Cathedral), Sabine, and Wheatstone.[8] Of these, all except Airy attended the January 1846 council meeting at which the committee was appointed, suggesting that Airy might have been co-opted on the initiative of his friends Herschel and Peacock. The committee's only recorded meeting took place at Kew Observatory on 7 May 1846, again suggesting little sense of urgency about the matter. The fact that Herschel had to ask Wheatstone for directions to the observatory emphasizes how little interest he had in the entire Kew project.[9] The presence of Airy on the committee would presumably have been a force against continuing the observatory, but the committee also had two of its staunchest advocates: Sabine and his ally Charles Wheatstone, who had strongly supported BAAS taking on the observatory in 1842.

According to the committee's report, signed by Herschel, it was "unanimously" agreed that the observatory at Kew should "be maintained in its present state of efficiency."[10] The surviving correspondence suggests that a report was written in advance by Wheatstone and then deliberated on at the meeting.[11] Wheatstone had also drafted BAAS's original 1842 prospectus for Kew, which had emphasized its use as an experimental station, so it may not be surprising that many of the reasons given in 1846 for keeping the observatory read almost like a repeat of those stated four years earlier for taking it on in the first place. The report emphasized the building's use as a convenient location for BAAS and as a repository for its instruments. The observatory was currently being used for "inquiries into the working of self-registering apparatus," which were now bearing fruit (see the final section of this chapter). Similarly fruitful was Ronalds's ongoing study of atmospheric electricity: the report noted that this had "in effect furnished the model of the processes conducted at the Royal Observatory." Moreover, the report noted that BAAS's occupancy of the building was at the Queen's pleasure and a sign of her interest in, and approval of, scientific research; if it were handed back now it might never again be available to science.[12] The report was accepted, with no recorded debate, by the BAAS Council at its meeting in London the next day, and the decision to keep Kew Observatory running was ratified at the annual meeting in Southampton the following September, when the observatory's annual £150 grant was also renewed.[13]

That Airy, as a member of the committee, agreed to keep Kew running might seem inconsistent with the idea that he saw Kew as competition to Greenwich and so wanted to see it closed. But the committee's report, rather in the same way that the 1842 prospectus had emphasized the usefulness of Kew as a place for experimental inquiries rather than as a permanent establishment, only recommended that the electrical observations and experiments with self-recording instruments be kept going. Almost nothing was said about the magnetic and meteorological work that was central to Sabine's own agenda. In this light Airy might have regarded Kew as a useful technical laboratory in support of Greenwich, with Ronalds giving invaluable advice on making and using apparatus for electrical observations. Indeed, two years later Ronalds claimed that Airy gave exactly this reason for agreeing to the continuation of the Kew observations in 1846: "That they should serve as Tests for newly invented meteorological Instruments & Experiments."[14] In any case, during 1845 and 1846, Airy had many other matters to worry about. He was heavily involved with work for the Railway Gauge Commission, set up to determine the standard gauge for Britain's rapidly expanding railway network; at the time of the meeting at Kew on 7 May, one of his Greenwich assistants was on trial at the Old Bailey for incest and murder; and in late September 1846, the Berlin astronomers Johann Galle and Heinrich D'Arrest announced the discovery of a new planet, Neptune, an announcement quickly followed by allegations of inaction on the part of Airy over predictions of the new planet's position sent to him by Cambridge mathematician John Couch Adams the previous year.[15] If Airy saw Kew as a problem at all at this stage, it would likely have been as but one problem among many.

Nothing further regarding the future of Kew Observatory appears on record for most of the next two years. Then, in April 1848 the BAAS Council asked the committee that had met in 1846 "to prepare a Report on what has since been done, and on the present state of the Observatory" in order to establish whether it was worth "continuing the present expenditure" on the observatory. This time the initiative seems to have come from the council and not the annual meeting, which was not held until August of that year. It is unlikely to have come from Airy, who was not present at the April meeting and was not a council member. The minutes do, however, record the co-option to the committee of Leonard Horner (1785–1864).[16] Horner's main interests were geology and the improvement of working-class education; he had also served as the most energetic of the four commissioners appointed to enforce the 1833 Factory Act, which attempted to limit the use of child labor in factories. Given BAAS's limited budget and the £150 a year that it was costing to run Kew Observatory, Horner might well have felt that the association had more urgent priorities.

The 1848 motion caused much more consternation than that of 1846, triggering a vigorous correspondence between Ronalds, Sabine, Herschel, and a meteorologist and astronomer who by 1848 was well known in the physical sciences community but had had little direct involvement with Kew up to then: William Radcliff Birt. Self-taught in the sciences (as far as is known), Birt was eagerly looking for a paid job in astronomy or meteorology. In 1842, soon after the news had appeared of BAAS's acquisition of Kew, Birt had sought a testimonial from Herschel in support of his application for the "curatorship" of the observatory.[17] In the late 1830s, Birt's work in both astronomy and meteorology had caught the eye of Herschel, especially his proposal for a long-term series of observations to detect "atmospheric waves"—pressure waves that some, including Herschel, believed might help explain the circulation of the atmosphere, something that was poorly understood at the time. Between 1839 and 1843 Herschel had supervised Birt in a project, supported by a £50 grant from BAAS, to reduce meteorological observations with a view to verifying the existence of these waves. This resulted in several papers in the BAAS annual reports in the 1840s and beyond. Vladimir Jankovic has shown how the theory of atmospheric waves became discredited after the late 1840s, when it gradually became apparent that there was no real evidence for the waves and the emphasis of meteorology had changed from theory-driven research to practical, utilitarian work in support of the navy and merchant shipping.[18] But in the 1830s and 1840s Birt's work impressed Herschel because it resonated exactly with Herschel's own approach to research: that data should be gathered for the purpose of putting theory to the test. Later, he was to praise Birt's analysis of the Kew observations of atmospheric electricity as an "interesting and thoroughly *inductive* discussion of a mass of obsns."[19]—words which might have been taken out of Herschel's *Preliminary Discourse.*

In May 1848, just six weeks after the future of Kew was again put under review, Birt wrote to Herschel to say that the possible impending closure of the observatory would terminate the five years' worth of electrical observations made there so far and that the work might therefore be for nothing if the observations were not continued and discussed "with the view of attempting the deduction of laws." Birt offered his services in continuing the electrical observations and running the self-recording instruments at Kew, as well as working on the reduction of the electrical results, "under the superintendence of Mr. Ronald [*sic*]."[20] Herschel agreed with Birt's rationale for keeping Kew Observatory going. He recommended to Sabine that in addition to continuing the observations there, "*Those already accumulated should be discussed with scientific precision*" (Herschel's emphasis) and that Birt's "liberal offer" of undertaking the work should be accepted. He acknowledged that the financial situation was difficult for BAAS, but that "the Association ought not, except on *very* urgent grounds, to throw up the observatory." It is a sign of the authority that Herschel commanded that the conclusion of his letter to Sabine reads almost like a military order: "Should it not be in my power to attend at the Council . . . you will oblige me by communicating this statement of my views . . . and by reading the letter enclosed."[21]

Sabine, however, was less sure about taking on Birt. For one thing, Sabine claimed that astronomy was Birt's real passion, much more than meteorology.[22] Moreover, at some point between 24 May and 17 June 1848, it became clear that Birt would require a salary of £100 per annum, thus putting an even greater strain on the observatory's tiny budget.[23] Ronalds hoped that Birt might take the place of the existing assistant, John Galloway, because Ronalds wanted a "properly qualified" observer who, in addition to routine reading of the instruments, could do the more complex work that now needed to be done, such as reducing the results.[24] But the proposed salary for Birt was nearly twice the £54 paid annually to Galloway. Little is known of Birt's personal life,[25] but he clearly had a different class and career background from the ex-army sergeant Galloway. The latter was expected to clean and maintain the building, so if he were replaced by Birt, someone else would have to be employed "to perform the menial duties of the house." Sabine, not surprisingly, preferred the existing arrangement of a "Servant" living on the premises and doing some basic observing as well as "work of a menial nature."[26]

The committee that had been reconvened in April 1848 met at Kew on 5 July and reported to the BAAS Council meeting two days later. At the end of June, Ronalds had sent Sabine a lengthy report, apparently at Sabine's request, on what had been achieved at Kew Observatory since 1846 and his views on its future, assuming that the committee and council voted to continue it. Ronalds emphasized the differences in the work at Kew from that at Greenwich—in particular, the "unique" observations of atmospheric electricity, which took a much wider variety of measurements than the Greenwich program, and Kew's far superior self-recording magnetic and meteorological instruments. If the observatory could not be kept running as it was (albeit with a better-qualified observer than Galloway), Ronalds asked that it at least be kept on as a depot for instruments. Failing this, he suggested, the association could give up the building entirely, "recommending that it may be supported on a sufficient Basis for using it as a Proving House &c . . . , by Her Majesty's Government."[27] By "proving house" Ronalds meant a place for testing and comparing meteorological instruments, something suggested in the original 1842 prospectus. We do not have a full report of the committee's 5 July meeting, but the minutes of the council meeting on 7 July suggest that the council took up Ronalds's recommendation for government support, for the committee was now asked to draw up a memorandum to the Treasury, asking "that means might be taken to preserve to the nation the benefit of the establishment of the Observatory at Kew." The memorandum was to state that the observatory's running costs, though not large, were beyond the means of BAAS and that the immense value of the work at Kew meant that there was a duty to maintain it.[28]

The idea that Kew Observatory should be a central, government-supported "proving house" may not originally have been Ronalds's. A week before Ronalds wrote his report, Sabine had suggested to Herschel that Kew could be turned into a "head quarter establishment" for instrument trials and comparisons as well as for magnetism and meteorology.[29] Herschel dismissed this idea at the 5 July committee meeting, but two weeks afterwards he confessed to Sabine that he may have been too "hasty" now that the proposal to apply to government had been taken up by the council. He now asked Sabine to draw up a draft of the proposed memorandum.[30] In his reply, Sabine wrote of the need for a government "establishment" for coordinating and reducing observations, due to a likely increase in the volume of meteorological and other observations coming in from the outposts of the British Empire.[31] Herschel's response began abruptly: "I cannot give my support to an application to Govt to take on itself the support of the Kew Observatory because I am not sufficiently impressed with the scientific necessity of such an establishment unconnected with the peculiar objects which made it desirable for the British Association as their private property." Herschel declined to attend any meeting for the purpose of drawing up a memorandum to the government.[32] John Cawood has claimed that by the 1840s, Herschel had become exasperated with what he saw as Sabine's obsession with data gathering, which conflicted with Herschel's theory-driven approach to all scientific enquiry.[33] He might have seen another example of this in Sabine's plan for a center that coordinated observations.

Herschel's views were reflected in the next report of what had now become known as the "Kew Observatory Committee." This was presented to the BAAS Council in August 1848, at the start of that year's BAAS annual meeting in Swansea. The report—which was signed by Herschel—claimed that the association could not continue the observatory even for another year on its current restricted budget and that to pursue "some of the most important objects which have all along been contemplated in its occupation"—including the standardization work outlined in the 1842 prospectus—would be quite beyond the means of the association. The report noted the possibility that the government would require some such central institution in the future but concluded that the committee saw no option but to discontinue Kew as soon as possible and seek "the most fitting mode of procedure for resigning it into the hands of Government."[34] The decision to discontinue the observatory was duly approved at the annual meeting, when the council was authorized to start closing it down. Birt was awarded a one-off grant of £50 for the reduction of the Kew electrical observations, but the observatory's annual grant was reduced to just £100, presumably for the purpose of winding it down.[35]

By January 1849 Galloway had been dismissed, but the council made no move at this or any other meeting to hand the building back to the government. Instead, the council resolved to continue the observatory "in its present state" until the next annual meeting in September, when the question of its continuance would again come under the scrutiny of the wider BAAS.[36] In any event, the 1849 annual meeting voted to continue Kew for another year, substantially increasing the grant to £250, though on the strict understanding that its continuation beyond the 1850 annual meeting was not guaranteed.[37] A letter from Lord

Northampton, now president of BAAS, reveals that the vote was passed with the intention of handing the observatory over to the government "in a year or two,"[38] so this was less of a change in policy from 1848 than at first appears. The decision to abandon Kew was not overturned in 1849: it was merely deferred. It is possible that the council members, despite the decision at the previous annual meeting, did not *want* to part with the observatory that had done, and was still doing, so much good work and had greatly added to the prestige of the association. In addition, the government's award, in the spring of 1849, of £250 to Francis Ronalds for his improvements to self-registering magnetic and meteorological instruments (described later in this chapter) must have further raised the observatory's profile and made it more difficult to put forth the case for closing it down. However, the documentary evidence suggests that it was Herschel who took the final decision to defer: the reason given for continuing the observatory, and increasing the grant, was that Herschel believed the Kew electrical observations to be "peculiarly valuable and likely to produce important results."[39]

In October 1849, the council agreed to appoint Birt to carry out observational work at Kew at a salary of £100 per annum. Again, Herschel seems to have been instrumental in the decision to take him on. His letter to Sabine one month before the 1849 annual meeting again has the tone almost of a command: "I should be very glad that anything should turn up by wh. Mr. B's . . . zeal for meteorl. obsn & reduction could be made available to Science. A continuance . . . of the Kew Electrical observations would no doubt be a desirable object, . . . '[40] Birt started work at Kew on 2 November. Yet almost from the start of his employment Birt seems to have taken on more than he had bargained for. Since the dismissal of Galloway, there had been nobody to do the laborious readings of the meteorological instruments as well as the electrical observations, which had to be made at regular intervals from early morning until late evening. This now fell to Birt, much to his chagrin. As early as 15 November he was complaining that he had been led to believe "that it was not at all contemplated to carry on a *regular* series of observations here but to attend more particularly to such objects as the [Kew Observatory] Committee . . . from time to time might determine on." From the start Birt also clashed with Ronalds, who, Birt claimed in the same letter, behaved in an "ungentlemanly" manner toward him.[41] This may have been partly a conflict of personalities, but in addition Birt did not seem to recognize Ronalds's authority as superintendent at Kew—something evidenced by Birt's correspondence, which describes his clashes with Ronalds in minute and sometimes remarkably petty detail. For example, Birt wrote that he was not allowed to alter the positions of any meteorological instruments in the observatory, despite none of them—according to Birt—being suitably positioned.[42] By late December, Birt was feeling that he had been employed at Kew as a servant, "in precisely the same capacity as Mr Galloway." The way he was treated seems to have caused him to have a nervous breakdown, and Birt was unable to continue with the electrical observations.[43]

The situation at Kew at the end of 1849 was made worse by Sabine being taken seriously ill in November of that year, possibly with some kind of fever, which made it necessary for all communications with Sabine to go via his wife, Elisabeth. The Kew Observatory Committee did not meet until 22 March 1850, when Sabine had recovered sufficiently. The committee directed that Birt was to make a reduced schedule of electrical observations three times a day for five days per week, together with meteorological observations at the same time.[44] Things did not improve for Birt, however, and on 5 June he informed Sabine that he would not be willing to work at Kew after the end of his first year there "*under present arrangements.*"[45] To Birt's horror, Sabine accepted this letter as Birt's resignation, with immediate effect from 5 June. In desperation, Birt wrote to his old mentor John Herschel, saying that Sabine had misinterpreted his letter. Herschel replied that he was "exceedingly sorry" about what had happened but, true to form, did not wish to become further involved. Shortly afterwards, Birt wrote that he had been refused an interview with Sabine and now acknowledged that his resignation was final.[46] He found little sympathy with other leading BAAS figures, due at least in part to Ronalds having friends in high places. In September, the geologist John Phillips, assistant secretary of BAAS, reflected that "Mr Birt *rues* as we say in Yorkshire of his unnecessary haste, but too late." Later that month, Phillips looked forward to going to Kew to "see my friend Ronalds again."[47]

Birt's disastrous time at Kew could well have stemmed from a notion he might have had that routine observing would be a secondary aspect of his work and that he would be able to devote most of his time at the observatory to research projects, such as his beloved atmospheric waves. Birt strongly suggested this in his letter to John Herschel: "The Association had entrusted to me the investigation and discussion of two very important subjects [analysis of atmospheric waves and electricity] in which as you are well aware I have been successful."[48] Not long after arriving at Kew, he had written to Herschel that he was thinking of applying to the Royal Society for a grant of £50 to support his research on atmospheric waves.[49] But perhaps the main reason why Birt was so unhappy at Kew was that he had hitherto done all his scientific work in his own time and—except for occasional payments from Herschel and BAAS—at his own expense. He had thus been free to pursue his own interests and choose a pattern of work that suited him. But as a paid employee of an observatory, reporting to a superintendent and a governing committee, Birt no longer had this freedom. His frustration in this regard is very apparent in his letter to Phillips on 15 November 1849, in which he complained about having to do "a *regular* series of observations" rather than one-off projects.[50] Birt had been taken on at Kew at the urging of John Herschel, who had admired Birt's research methods because they appealed to Herschel's theory-driven approach. But Herschel's approach was not what was required at Kew. Sabine required a loyal subordinate in his ranks who would dutifully take the data that Sabine wanted. Birt was not such a person.

Sabine and his colleagues on the Kew Observatory Committee, known from May 1850 onward as the Kew Committee, would have had fewer doubts about Birt's successor. John Welsh (figure 2.1) was born into a middle-class family in southwest Scotland and educated at Edinburgh University—in part under James Forbes, one of the instigators of the Royal Society's original 1840 attempt to establish a magnetic and meteorological observatory.[51] Unlike Birt, Welsh was used to working as part of a team in a highly disciplined environment, and he also had much experience of the type of observational work being done at Kew. Since 1842 he had worked at the magnetic and meteorological observatory at Makerstoun, Scotland, run by Sir Thomas Brisbane, a former soldier and governor of the penal colony of New South Wales, where he had founded the Paramatta Observatory. Brisbane was a patriarchal figure who ran observatories rather like the colony.[52] According to an anonymously written obituary, Welsh's appointment to Kew owed much to the then chairman of the Kew Committee, William Henry Sykes, to whom he had been recommended by Brisbane and John Allan Broun, Welsh's immediate superior at Makerstoun.[53] Sykes had spent most of his working life with the East India Company; like Brisbane and Sabine, he was an army officer who had had scientific roles, in Sykes's case compiling statistics on British India. The obituary's claim is believable: Brisbane's

FIGURE 2.1. John Welsh FRS (1824–1859), superintendent of Kew Observatory, 1852–1859. Image courtesy Met Office National Meteorological Library and Archive.

recommendation would have been received sympathetically by his fellow soldiers, Sabine and Sykes. In addition, the timing of Welsh's availability was convenient, for in 1850 the magnetic and meteorological work at Makerstoun was closed down and Welsh was made redundant. On 5 July 1850, less than a month after Sabine took Birt's expression of dissatisfaction as his resignation from his post, the Kew Committee decided to employ Welsh at Kew.[54]

Just 25 years old at the time of his appointment, Welsh immediately settled into his new job. Subsequent Kew Committee reports are almost gushing in their praise of Welsh: "The zeal and intelligence with which Mr. Welsh has continued to execute his duties has given the Committee unmixed satisfaction."[55] Ronalds remained at Kew as honorary superintendent, though he resigned in late 1852, after which Welsh took over the running of the observatory. Neither the BAAS Council minutes nor the minutes of the Kew Committee contain any record of Ronalds's resignation or his reasons for leaving. His resignation might well be connected with the death of his mother in 1852 (Ronalds never married) and his acceptance of a Civil List pension in honor of his scientific work.[56] However, probably just as important was the fact that, as described later in this chapter, by 1852 the work at Kew was very different from what it had been when Ronalds had arrived a decade earlier. Ronalds was fundamentally an inventor and designer of new instruments, but by the early 1850s, making and testing commercial meteorological and magnetic instruments on a routine basis had become central to the Kew regime. This standardization work, as Ronalds himself confessed a few years later to Armagh Observatory director Thomas Romney Robinson, was not to Ronalds's liking.[57]

Although the observatory's long-term future was by no means secure, in October 1849 the committee originally appointed in 1846 now effectively became permanent. The committee's brief was now "visiting and exercising a general superintendence" over the activities at Kew.[58] Welsh was appointed as the "observer at Kew,"[59] a title reminiscent of that of "King's Observer," used to describe the director of the observatory when it was George III's private establishment. Such language, like the phrase "Kew Committee," is indicative of the observatory's increasing prestige within BAAS. More and more it was being thought of as a permanent institution, even though funding was not guaranteed beyond the 1850 annual meeting. In addition, from mid-1850 onward, Kew Observatory started to become more like a business. Beginning June 1850, formal minutes were kept of Kew Committee meetings. On 5 July, the Kew Committee decided that the observatory's complement of staff "should consist of an Observer and a Mechanic"; at the same meeting, Ronalds reported that he had engaged Mr [Richard] Nicklin, whom he described as a "photographic mechanic."[60] Thus from mid-1850, Kew had a scientifically qualified "observer" (Welsh), with Nicklin assisting him with the mechanical work. Starting on 27 August, Ronalds and Welsh kept a diary of all activities at Kew, including visitors and administrative changes. The diary was maintained until 31 October 1851.[61]

It is possible that this new businesslike nature of Kew Observatory was connected with John Herschel's appointment, in December 1850, to the post of master of the Royal Mint.[62] The position left Herschel with little time for scientific pursuits, and from then on his direct involvement with Kew Observatory ceased. In 1849, the authority commanded by Herschel had led to the appointment of Birt, against the wishes of Sabine. Now, with Herschel no longer playing a leading role, Sabine had a free hand to appoint loyal subordinates and set his own agenda at Kew. However, a more immediate reason is likely to have been the appointment of John Peter Gassiot to the Kew Committee in October 1849. A Fellow of the Royal Society since 1840, Gassiot was respected as a chemist and was renowned for his spectacular electrical experiments at his London home. Most importantly for the subsequent history of Kew Observatory, Gassiot was a businessman through and through who had made his fortune as an importer of port wine. The new regime at Kew, with its division of labor into "observer" and "mechanic," directed by a permanent committee that kept regular minutes of its meetings, shared several characteristics with those of a business concern. Gassiot's future correspondence would emphasize this change of regime, as would his role in the introduction of instrument standardization at Kew.

At the 1850 annual meeting of BAAS, Welsh's salary was still only guaranteed for the following year. At the same meeting, the BAAS General Committee asked the council to contact the Royal Society—and, if need be, the government—as to "the possibility of relieving the Association from the expense of maintaining the establishment at Kew."[63] But this was to be the last grumble about the cost of maintaining Kew to appear in the BAAS Council minutes for many years, for by the time of the 1850 annual meeting, the observatory's finances were improving. At this meeting, the BAAS annual grant to Kew was increased to £300; according to Ronalds, this was due mostly to unrecorded behind-the-scenes actions by John Gassiot.[64] By the end of the 1850s, the BAAS grant to Kew had increased to £500.[65]

That there was any difficulty at all about the grant may have been due to an older enemy, George Airy, who at this meeting became BAAS president for the 1850–1851 session. Just days after taking office he wrote to Kew Committee chairman William Sykes, expressing his view that the long-term observations currently in progress at Kew should be terminated because Airy believed that the original purpose of Kew was the testing of newly invented instruments, not continuous observations. Airy also remarked that Kew would not obtain any government support for such continuous observations.[66] As in 1840, Airy seems to have been concerned that Kew was duplicating the regular magnetic (and now meteorological) observations that were already receiving government support at Greenwich. In his reply, sent with the approval of the Kew Committee, Sykes assured Airy that the observatory's primary purpose, that of experiment, was always kept in view and that such long-term observations as were in progress were all for specific purposes in that the barometers, magnetometers, and other instruments all required long periods of observational testing to be verified.[67] Airy made no further recorded moves against Kew for the rest of his presidency of BAAS. We do not know whether he really believed Sykes's response that the data gathering at Kew was secondary to its main purpose of instrument testing, but he must have realized that there was no moving the Kew Committee.

The 1850 grant increase shortly followed another turn of events which, together with the enhanced BAAS money, would "render 'poor Kew' rich," as Ronalds put it in a letter to Sabine.[68] In late 1849, there came the announcement that the Whig government of Lord John Russell intended to provide an annual grant of £1,000 to the Royal Society for scientific purposes. Although there was no guarantee that this new "government grant" would be permanent—indeed, it was nearly terminated when Lord Palmerston became prime minister in 1855—from the beginning the Royal Society worked on the principle that it would last indefinitely.[69] Given that active members of the Royal Society Council often played prominent roles in BAAS as well, it was not long before some leading BAAS figures saw the potential of the government grant for supporting Kew Observatory. Sabine became secretary of the Royal Society's new Government Grant Committee, while Murchison, an enthusiast for Kew Observatory since 1842 and ever keen to seek influence with the aristocracy and government, became its chairman. Murchison, in particular, was ecstatic about the prospect of government money for scientific research, dubbing it in March 1850 "the California of the Government thousand"[70]—a reference to the California gold rush of the previous year.

Soon after the government grant was publicly announced, Murchison, wanting to keep the observatory running *"coûte qu'il coûte,"* suggested to Herschel that with this new source of government money, the Royal Society might take responsibility for Kew Observatory if BAAS had to give it up. Also, he agreed with Sabine "that a good national Physical Observatory should be sustained at Kew."[71] Once again, Sabine was advocating a state-supported observatory but once again, Herschel refused to give his backing. Herschel believed that for the Royal Society to commit itself to the maintenance of any observatory or to spending any portion of a grant for an indefinite period of years "should most earnestly deprecate the RS." This does not contradict Herschel's earlier advocacy of state-funded physical observatories, for he had always believed that these should be run directly by the government and not scientific societies. For Herschel, the permanent maintenance of observatories and laboratories was not part of the Royal Society's mission. Herschel was no less enthusiastic than Murchison about the new grant, which he saw "as a Godsend to British Science." But Herschel believed that it should be used, first, to assist *"Private individual Experimental Research"* (Herschel's emphasis); second, for analysis and reduction of observations already made; and also for occasional, undefined, special scientific projects of fixed duration.[72] Herschel therefore saw the grant very much as an extension of the existing culture in which scientific research was funded primarily by private individuals of independent means.

Murchison, when he presented the report of the Government Grant Committee to the Royal Society Council, deferred to Herschel's views on how the grant should be distributed,[73] and so once again Herschel punctured the idea of Kew Observatory being funded by the state. However, a major success for Kew came with the announcement of the very first round of awards out of the government grant in 1850, when £100 was awarded to Sabine for new instruments at Kew Observatory. The news may not have been a great surprise, given that the committee making the awards had two of the observatory's loudest advocates as its chairman and secretary, while a third staunch supporter, Wheatstone, was on the subcommittee recommending the award. The money was spent on a new vertical force magnetograph—which helped put the Kew magnetic observations on a well-funded, permanent footing—as well as modifications to a Daniell hygrometer and also a "standard thermometer" with which other thermometers could be compared.[74] Although no proposal was ever put forward in these years for the Royal Society to maintain Kew Observatory out of the government grant, the Kew Committee continued to be successful in attracting substantial grant income from the Royal Society. Out of the 1851 government grant, £150 was awarded for the construction and verification of standard meteorological instruments at Kew, as well as for the purchase of an apparatus for graduating thermometer tubes. The money provided by both the 1850 and 1851 grants was thus important in establishing a program of instrument standardization at Kew (discussed in the next section). An additional award of £175 was made to George Gabriel Stokes, Lucasian professor of mathematics at Cambridge, for experiments to be carried out at Kew to determine the indices of friction of various gases.[75]

That members of the Kew Committee so quickly, and successfully, applied for money from the government grant also raises the possibility that they might have been anticipating the announcement of such a grant. It is interesting that despite a very firm statement in August 1848 that the BAAS Council would close the observatory and give the building back to the government, in fact it never quite got around to doing so. Instead, the council deferred the issue until the 1849 annual meeting, when the observatory was reprieved for another year until 1850. The origins of the government grant are somewhat obscure, Prime Minister Lord Russell's initial letter to Lord Rosse, allegedly dated 24 October 1849, having long since disappeared.[76] It is known, however, that John Herschel was a friend of Lord Russell. Furthermore, in March 1849 Herschel and Lord Northampton went to see the prime minister to seek financial reward for some of Francis Ronalds's inventions (see the final section of this chapter).[77] This leaves open the possibility that the grant might have been the result of unrecorded discussions between these three. That Herschel might have been anticipating the announcement of the grant is made more plausible by the fact that he was chairman of the Kew Committee in 1849 when it agreed to defer closing the observatory. More particularly, the resolution at the 1849 BAAS annual meeting, to keep Kew Observatory running and to increase its annual grant from BAAS, was cited as being in response to John Herschel's favorable opinion of the ongoing electrical observations at Kew. This resolution is dated 19 September 1849—barely more than a month before the date of Lord Russell's alleged letter and less than two months before the earliest recorded minutes of the Royal Society's Government Grant Committee.[78]

Despite the presence of Sabine and Murchison on the Government Grant Committee, Kew had to compete with a substantial, and growing, body of applicants from other sciences. Not all of the money awarded to Kew by the Royal Society came from the government grant. For example, the £261 awarded in 1852 for a series of meteorological balloon ascents under the direction of the Kew Committee was from the Society's Wollaston Donation Fund—an older, private source of funding.[79] In an environment in which government support for science remained very limited—in addition to the BAAS grant to Kew remaining very modest—the Kew Committee had to look for other sources of income. It was in this context that, in the early 1850s, the Kew Committee began pursuing an enterprise that would soon be self-financing and would bring the observatory to the notice of a far wider circle than hitherto: that of instrument standardization.

THE ORIGINS OF STANDARDIZATION AT KEW OBSERVATORY

The introduction of standardization at Kew—in the form of making standardized instruments or testing instruments sent to Kew by external makers—was part of the new businesslike regime implemented at the observatory by Gassiot and Sabine from mid-1850 onward. A form of standardization on a small scale had been one of the objectives listed in BAAS's 1842 prospectus, which suggested that Kew should act as "a station to which persons . . . may bring their instruments for the purpose of comparison with the standard instruments there deposited."[80] However, this was not implemented immediately, possibly due to the observatory's very restricted budget for most of the 1840s, in addition to Sabine's first priority being to establish Kew as an independent magnetic and meteorological observatory.

Standardization at Kew began in September 1850, when the newly recruited John Welsh began experiments to compare hygrometers made by John Frederic Daniell and the well-known French chemist and instrument maker Henri Victor Regnault.[81] By 1850, Regnault was well known for his work on the physical properties of the steam engine. He was also a leader in the field of precision measurement, greatly improving the accuracy of thermometers. One of his hallmarks was the careful elimination of errors during the measurement process itself, rather than simply correcting them afterward. He was highly respected by men of science across Europe, including James Forbes and the youthful William Thomson.[82] In November 1850, the Kew Committee used its award from the first Royal Society government grant to purchase from Regnault "a standard thermometer . . . every degree of which shall have been examined and shall be guaranteed by M. Regnault himself" in order to verify or correct thermometers made by British instrument makers.[83] By early 1851, it was clear that the Kew Committee's ambitions went further: in addition to verifying thermometers, the committee now also proposed to use the Regnault thermometer as a standard for making thermometers, using a graduation apparatus by French engineer and inventor Louis-Guillaume Perreaux. The machine was initially paid for by Gassiot and arrived at Kew in February 1851. The BAAS Council minutes claimed that it was obtained via Gassiot's actions "in anticipation" of money being received from the 1851 government grant.[84] In July, Regnault himself visited Kew, advising on the use of the Perreaux dividing engine and on his method of calibrating, graduating, and testing thermometers. On Regnault's suggestion, the Kew Committee invited Perreaux to Kew to iron out some technical problems with the machine; Perreaux visited Kew in October 1851.[85]

Sabine also took a leading part in this initiative. In October 1850, he wrote to Ronalds that "my head is full of apparatus for making standard Meteorological Instruments of different kinds—we ourselves at Kew to be the makers; & the instruments to bear our stamp."[86] Five months later, just days after the Perreaux machine had been put into operation, Sabine "suggested the desireability [sic] of dividing thermometers at once with Fahrenheits degrees instead of an arbitrary scale."[87] Further evidence of Sabine's leading role can be found in his friendly correspondence with John Welsh, who rapidly took charge of the standardization program at Kew: as early as April 1851, the Kew Committee claimed him to be a "master" in the use of the graduation machine. Sabine came to treat Welsh as a personal friend, as is exemplified by his letter offering Welsh "a card for Lord Rosse's Soirées." Welsh duly complied with Sabine's call for Fahrenheit-scale thermometers: his letter to Sabine dated 17 December 1851 encloses a step-by-step account of a process he invented for graduating a thermometer in degrees Fahrenheit.[88] In January 1852 the Kew Committee reported that the thermometers made by Welsh, when compared with each other and with the Regnault standard, had been found to be "highly satisfactory" by the Royal Society's Government Grant Committee and that "standard instruments bearing the mark of having been constructed and verified at the Kew Observatory" had been supplied to the imperial observatory at the Cape of Good Hope. Further thermometers had been ordered by the Hobarton observatory in Tasmania (part of the Magnetic Crusade) and by James Forbes for his experiments on heat.[89]

Sabine's fellow scientific servicemen also played an important role in instrument standardization. In January 1852 the Kew Committee reported that John Welsh had begun some "very laborious verifications" of twenty sets of meteorological instruments to be used by the East India Company for meteorological observations in India.[90] Until 1858, the East India Company managed Britain's imperial possessions in India, including the military regiments. The demand for instruments is likely to have been in response to a program of meteorological observations across the British Empire instigated by army officer William Reid, who had long been interested in the causes of tropical storms. This possibility is much strengthened by the fact that Reid briefly served on the Kew Committee and BAAS Council in the early 1850s.[91] But the person most likely to have put this new East India Company initiative into effect was the Kew Committee chairman, William Sykes, who was one of the company's directors. Testing large numbers of thermometers soon became an ongoing practice. In March 1852, the BAAS Council authorized the Kew Committee "to supply Standard Thermometers, on official application, to any department of Her Majesty's Government or the East India Company." Many of these thermometers were not made at Kew: they were manufactured by London instrument makers and then tested at Kew. That Sabine was fully behind the mass production of thermometers for the East India Company and other bodies is attested to in his letter to Welsh one day before the March BAAS Council meeting: "It is extremely desirable that we should meet the applications from Government as far as we may be able to do so."[92] By late September 1854, some forty thermometers, twenty hygrometers, and four barometers for the East India Company had been tested at Kew, as had several more meteorological instruments for other institutions. By the same date, a total of ninety-four thermometers had been made at Kew for "institutions and individuals."[93] Thus by the mid-1850s, Kew had become an imperial capital of meteorological instruments, a place that the British Empire relied on for the manufacture and testing of these instruments. Alex Soojung-Kim Pang has shown how technology, components, and observation methods developed in the mother country were necessarily "hard-wired into" astronomical instruments used at imperial observatories and on eclipse expeditions.[94] In the same way, the technologies developed at Kew were becoming essential to the British Empire's meteorological needs.

The standardization work at Kew soon assumed a commercial aspect. At the same 1852 meeting it authorized the Kew Committee to supply thermometers to the government and the East India Company, the BAAS Council sanctioned the sale of Kew thermometers to individual BAAS members or Fellows of the Royal Society at £1 per instrument and also their manufacture for "certain of the Philosophical Instrument Makers."[95] Then, the 1853 International Meteorological Conference and the consequent establishment of the Meteorological Department of the Board of Trade had the effect of further expanding the role of Kew Observatory as a standardization center because it obliged British ships to provide weather reports to an internationally agreed-upon standard using standardized instruments. By the end of the conference in early September 1853, the Kew Committee had agreed to provide Admiralty hydrographer Francis Beaufort with a specimen of a thermometer specially adapted for meteorological observations at sea, and Welsh was set to work on constructing one.[96] Welsh must have completed the task very quickly, for on 3 December Beaufort was informed that samples of the Kew marine thermometer had been sent to various London instrument makers with requests for the prices at which they could supply such thermometers in bulk. Two well-known firms, Casella and Negretti and Zambra, were selected as having quoted the lowest prices, and Beaufort was informed that either of these companies could supply the Admiralty with thermometers for just five shillings sixpence (5s 6d) apiece. In April 1854, Kew entered into a similar agreement with the Board of Trade to provide meteorological instruments for merchant shipping.[97] By mid-1855, the observatory had tested the accuracy of more than 2,000 thermometers; of these, 400 were for the Admiralty and 480 for the Board of Trade.[98]

These mass-production thermometers were all made by the instrument makers and then tested at Kew. The two companies, Casella and Negretti and Zambra, had been founded earlier in the nineteenth century by Italian immigrants and both already enjoyed a fine reputation. In an 1860 advertisement, Casella proudly listed Kew as one of the "Royal Observatories" that it supplied its wares to.[99] In the sense of the instruments being verified at Kew Observatory before being sent on to the Admiralty and Board of Trade, Kew Observatory can be considered the customer for as well as the inventor of the marine thermometers made by these firms, anticipating a tradition of partnerships between maker and customer-inventor identified by Mari Williams as beginning much later in the nineteenth century.[100]

Fully half of the 2,000 thermometer tests reported in mid-1855 were for the United States government, not the British. In August 1853 Maury, while in Europe for the meteorological conference, met with Sabine and informed him that he was not satisfied with the marine barometer then being used by the US Navy. Through Sabine, he obtained the agreement of the Kew Committee to make a better one.[101] Kew had been verifying barometers—and also hygrometers for measuring atmospheric humidity—since late 1852 for the East India Company,[102] but making a barometer for use aboard ship posed a special problem due to errors in the level of mercury in the tube caused by the motion of the vessel at sea. Welsh and London instrument maker Patrick Adie corrected the problem by suspending the barometer freely on an arm attached to the wall of a ship's cabin and by constricting the lower part of the tube. Welsh and Adie's design proved successful in sea trials, and the barometer adopted by the Kew Committee in March 1854 became known as the "Kew pattern" or "Kew type" barometer; Negretti and Zambra referred to it as the "Kew Marine" barometer.[103] This model of barometer was selected to be supplied to the US Navy. By mid-1855, some fifty marine barometers of this model had been dispatched from Kew to the United States. At the same time, Kew also sent a thousand verified thermometers to the US Navy.[104]

Apart from the requirements laid down at the meteorological conference, it should not be surprising that Maury wanted to have these new marine thermometers and barometers all calibrated to the highest standard available. The US Naval Observatory, of which Maury was superintendent, was keen to catch up with its European counterparts and had by now established a practice of touring Europe for the best instruments. Already in the 1840s, its astronomer James Melville Gillis had visited some leading European observatories and instrument makers to procure the best astronomical equipment for the new establishment on the other side of the Atlantic.[105] These visits by Maury and Gillis also fitted in well with the strong economic links that prevailed between Britain and the United States in the period from the conclu-

TABLE 2.1. Kew instrument verifications, 1854–1859

	1854–1855	1855–1856	1856–1857	1857–1858	1858–1859	TOTAL
Thermometers	2520	530	1524	268	911	5753
Barometers	257	137	278	221	187	1080
Hydrometers	1269	100	751	150	92	2362
TOTAL	4046	767	2553	639	1190	9195

Source: Data from BAAS:CM: 12 September 1855, 6 August 1856, 26 August 1857, 22 September 1858, and 14 September 1859.

sion of the War of 1812 (in which Sabine had served) and the start of the American Civil War in 1861—in spite of lingering political tensions between the two nations.[106] Yet instrument verifications at Kew for foreign governments did not stop with the United States: for example, twelve barometers were verified at Kew for the Portuguese government in the 1855–1856 period.[107]

Also during 1855, Welsh completed setting up a standard barometer at Kew with which other barometers could be compared. The Kew Committee soon felt confident enough to arrange with instrument makers to make Kew-verified barometers and thermometers available to the public at prices of £4 4 shillings per barometer and £2 2 shillings for a set of six thermometers, with advertisements to be placed in newspapers to this effect. The statistics for meteorological instruments verified at Kew in the 1850s are remarkable: between mid-1854 and mid-1859, the observatory tested more than 9,000 instruments for British and foreign government departments, instrument makers, and private individuals (see table 2.1). Importantly, this standardization work was profitable. The income from verifications was such that once they had covered the cost of the barometer verification apparatus, the Kew Committee felt able to reduce the charge for testing barometers from ten to five shillings per instrument.[108] By the late 1850s, testing instruments was bringing the observatory around £100 per year.[109]

These revenues made a notable difference to the observatory's annual income, as its annual grant from the BAAS Council was still only £350 for the year ending August 1857.[110] Thus a major—and publicly visible—part of the observatory's work was effectively self-financing. Moreover, it was also an essential government service, which made it much harder to make any case for closing down the observatory. This adds further weight to the idea that the introduction of standardization in 1850 was an astute long-term move by Gassiot and Sabine. Gassiot, ever the entrepreneur, likely saw the commercial potential of the standardization work and collaborated with Sabine to instigate it—even to the extent of paying for the Perreaux dividing engine out of his own considerable pocket. Gassiot succeeded Sykes as chairman of the Kew Committee in May 1853[111]—and would remain in this office until 1871. Perhaps even more importantly, Kew Observatory had gained considerable public authority as a center for top-quality meteorological instruments—overseas as well as within Britain. This is attested to by Casella mentioning Kew in its advertisement, in addition to Kew testing instruments for government departments. Although it may be going too far to claim that by the early 1850s Kew was "the acknowledged source of thermometers and barometers for expert observation,"[112] Kew was well on the way to being able to make this claim by the end of that decade.

It may be asked why all this large-scale standardization work went to Kew and not to Airy's Greenwich, which at the time had an established program of rating chronometers for the Admiralty and so might have been a natural first choice for testing meteorological instruments when the need arose. Some thermometers used by volunteers in the British Meteorological Society and some barometers destined for ports were, indeed, verified at Greenwich under James Glaisher,[113] but these were an exception to the general rule, as in the 1850s the great expansion of weather recording initiated by the Brussels conference applied only to marine meteorology, not land stations like Greenwich. There are several possible reasons for the dominance of Kew in this field. First, it is evident that two key members of the Kew Committee—Sabine and Sykes—had close connections with the Admiralty and the East India Company, two major customers for the standardization work.[114] Perhaps more important was that with the instigation of the thermometry program in 1850, standardization was established at Kew independent of government, before the demand came from the East India Company, the Admiralty, and the Board of Trade, thus leaving Airy and Greenwich out of the loop. Indeed, Airy was an early customer for one of the Kew thermometers.[115] In any case, Airy might not have been interested in entering this field for himself. Jim Bennett has described how Airy, despite having a lifelong interest in the science of horology and the improvement of chronometers, took objection to staff time at Greenwich being taken up with routine work on chronometers.[116] This suggests that there were limits as to what Airy regarded as his territory.

METEOROLOGY AND GEOMAGNETISM AT KEW, 1845–1859

Alongside the new, high-profile work in standardization, meteorological observation at Kew during the 1845–1859 period developed enormously from the initial basic program of observations with manually read instruments that had been established in the mid-1840s. In particular, these years saw the development of new types of meteorological instruments. At the same time, Sabine succeeded in establishing his cherished magnetic observations at Kew, again using innovative self-recording instruments. Although funding for the Magnetic Crusade ended in 1848, long-term magnetic observations continued to be carried out under Sabine's direction at several colonial observatories as well as at Kew. It is possible to argue that through Kew Observatory, Sabine managed to keep a Magnetic Crusade of sorts going throughout the 1850s and beyond. Both the magnetic and meteorological work at Kew did at least as much as standardization to raise the observatory's national and international prestige so that by the end of the 1850s Kew was renowned as a world center for research in both branches of science.

Observations of atmospheric electricity continued after the appointment of John Welsh in 1850, though thereafter they were mentioned less in BAAS reports, suggesting that the program had become a watching brief rather than frontline research.[117] In 1856, the electrometer assembly was removed from the observatory dome to make way for the solar telescope (see chapter 3), and a smaller electrical apparatus was built on the side of the dome.[118] Yet Ronalds's original apparatus remained a prototype for other observatories: copies were built at Kew for the East India Company in 1847 and the Madrid Observatory in 1852, suggesting that Kew was respected as an international standard in this field.[119] A new type of instrument for observing atmospheric electricity, devised by William Thomson, would be installed at Kew in the 1860s (see chapter 5).

Work on geomagnetism came to Kew in the mid-1840s with a project by Ronalds to build a self-recording instrument for measuring magnetic declination—the difference between true north and magnetic north. This "declination magnetograph" was an extension of the 1840s trend toward instrument automation, which made use of the new technologies of photography and telegraphy. The device worked using a magnet with a mirror that reflected light through condensing lenses to form a concentrated spot of light on a moving strip of photographic paper, thus making a trace of the variations in the magnetic field over time. The magnetograph required a vibration-free support. This was achieved by suspending it between two masonry pillars that had originally supported the transit instrument used to determine time for King George III's clocks—further evidence that in the Kew building, BAAS had a ready-made environment for precision measurement.

Ronalds published the details of his invention in *Philosophical Transactions of the Royal Society*.[120] The same volume contains a description of a very similar device built by Charles Brooke, a surgeon who invented a number of scientific instruments in his spare time. By mid-1847, Brooke was automating the magnetic and meteorological instruments at Greenwich; a year later these were up and running.[121] On the recommendation of George Airy, the Admiralty awarded Brooke a prize of £500 for his invention. Nothing, however, was awarded to Ronalds, even though Ronalds published his work at the same time—and despite some evidence that Airy was aware that Ronalds had been working on similar devices,[122] suggesting a further example of Airy's antipathy to Kew. Lord Northampton and John Herschel then applied to the government to seek some recognition for Ronalds. This led to Northampton and Herschel having a meeting with the prime minister, Lord Russell, in March 1849. The following month, Ronalds was awarded £250; the money was apparently forwarded directly from the prime minister.[123] This successful application by Northampton and Herschel must have brought the work at Kew to the notice of Lord Russell's government and might possibly have made that government more receptive to the idea that it could or even should reward scientific research with public money. Therefore, it may have been more than a coincidence that just six months later Russell, according to later accounts, wrote to the Royal Society's president, Lord Rosse, with the original offer of an annual £1,000 government grant for scientific research. As argued earlier in this chapter, this suggests that the Kew Committee, acting on Herschel's advice, might have deferred closing the observatory in anticipation of government funding.

In April 1851, Ronalds began a six-month trial of two further magnetographs that he had designed, along with the original declination magnetograph, forming a suite of instruments that recorded the three essential elements of Earth's magnetic field: declination, intensity, and dip (deviation from the horizontal).[124] The trial was completed and the results reported, but these instruments do not seem to have been used on a permanent basis during the early 1850s. The Kew Committee claimed as reasons limited funds and the priority of experiment over long-term observation at Kew. Moreover, the work was funded by a one-off grant of £100 from the Royal Society's donation fund.[125] Certainly the funding situation was not generous in the early 1850s, but the introduction of standardization must also have consumed much of the limited time and resources available, especially as Welsh had only one assistant in the observatory at this time. Yet there is evidence that some magnetic observations were being carried on quietly at Kew. In March 1850, Sabine obtained authorization from the BAAS Council to commence a modest program of magnetic observations at Kew, to be made one day per week using manually read instruments in a portable wooden observatory set up in the observatory grounds (to isolate the instruments from iron).[126] Both the instruments and the hut were "lent' by Sabine, suggesting that they were hardware left over from the Magnetic Crusade. Correspondence between Welsh and Sabine further demonstrates that some magnetic work was being kept going—as in a comparison by Welsh of the magnetic dip at Kew with that at Woolwich.[127]

During these years, equipment for other observatories and expeditions was also made at Kew. In 1850, Ronalds completed for the Toronto observatory (one of the stations that had participated in the Magnetic Crusade) an instrument for measuring magnetic dip. In early 1858, magnetic instruments were made at Kew for David Livingstone's upcoming expedition to Africa, and members of Livingstone's party came to Kew to be trained in their use—an example of the observatory acting as an imperial capital for magnetic instruments as well as meteorological ones.[128] Indeed, through this system of supplying instrumentation and training to overseas observatories and expeditions, in addition to quietly carrying on a low-level program of geomagnetic observations, some of the work of the Magnetic Crusade continued at Kew even after government funding for it had officially ended. A permanent program of magnetic observations with self-recording equipment began in January 1858 with improved versions of Ronalds's instruments, built with the aid of £250 from the Royal Society government grant. These observations were in response to Sabine's 1852 discovery of a correlation between sunspot activity and variations in Earth's magnetic field and were designed to run in parallel with a program of solar photography at Kew.[129] With this new set of self-recording instruments, Sabine at last had a magnetic observatory that could rival Greenwich.

In 1855, Kew Observatory was given international promotion when a selection of Kew magnetic and meteorological instruments was displayed at that year's Universal Exposition in Paris. This international exhibition, which ran from May to November 1855, was a deliberate attempt by Emperor Napoleon III to outshine Britain's 1851 Great Exhibition and flaunt the glory of the new Second French Empire, Napoleon III having become emperor only in 1852 after several years of political turbulence in France.[130] As in 1851, a major aim of the Paris exhibition was to display the finest examples of scientific progress from around the world. The Paris exhibition would be viewed by a huge audience, so it was a great coup for Kew to have instruments exhibited there. The involvement of Kew Observatory with the Paris exhibition originated with a letter sent in December 1854 from the Board of Trade to Lord Wrottesley, Lord Rosse's successor as president of the Royal Society. The letter claimed that the manufacturers of some scientific instruments, including "those for conducting researches into the laws of magnetism, heat, light, electricity, and other physical forces," were underrepresented in the list of British manufacturers proposing to send displays to the exhibition. The Board of Trade asked whether the Royal Society could appoint a committee to cooperate with the board in filling this deficiency. The committee appointed by the Royal Society included three important members of the Kew Committee: Sabine (now treasurer of the Royal Society), Wheatstone, and Warren De La Rue, who was starting to play an important role at Kew with the establishment of solar photography there. It may not be surprising, therefore, that at the next meeting of the Kew Committee, Sabine announced that the Royal Society would be happy to accept the help of the Kew Committee in sending a display of magnetic and meteorological instruments to Paris. The Kew Committee appointed John Welsh and mechanical assistant Robert Beckley to accompany the instruments to the French capital.[131]

One might have expected Greenwich, Britain's premier government observatory, to have a display at this exhibition. Yet it was Kew and not Greenwich that took the 25-foot-long exhibition space in Paris, alongside another 25 feet of counter space for smaller instruments. Both spaces were largely taken up by Kew apparatus, such as a self-recording magnetograph by Francis Ronalds, and the thermometer-testing apparatus. Some Greenwich instruments were exhibited, but they were only brought to Paris as a small part of the Kew display. The Board of Trade refunded the expenses, totaling just over £200, incurred in transporting the instruments to Paris.[132] The Board of Trade might have had good reason to be grateful to the Kew Committee and to promote the work of the observatory, given the huge number of meteorological instruments now being tested at Kew for British government departments. Interest in Ronalds's instruments was such that it led Ronalds—now retired from Kew and living in Paris—to reprint his BAAS reports as a short book.[133] The Paris exhibition symbolizes how, by the mid-1850s, Kew was becoming nationally and internationally recognized as an important center for designing, building, and testing magnetic and meteorological instruments.

The regular meteorological observations with manually read instruments were continued at Kew throughout the 1845–1859 period, mostly without interruption. At around the same time he invented the self-recording declination magnetograph, Ronalds also developed instruments that automatically recorded temperature, barometric pressure, and atmospheric electricity at Kew.[134] In 1854, John Welsh designed and built a new type of screen for shading thermometers from direct sunlight. This was an early form of "louvered" thermometer screens that allowed air to circulate freely around the thermometers while keeping them shaded.[135]

One enterprise that might have brought Kew to wider public notice was a series of four balloon ascents in 1852 to make meteorological observations at high altitude. These ascents were spectacular public events: they took place in the large *Nassau* balloon owned and operated by Charles Green, the best-known balloon pilot, or "aeronaut," of the time, and the choice of London's Vauxhall pleasure gardens as a launch site meant that they would have been noticed by many people. Yet the very appeal to the public of balloon launches as a spectacle and social event meant that ballooning was not seen as a serious activity,[136] which may explain why Welsh's ascents were the first serious scientific balloon flights in Britain. We do not know what motivated BAAS to enter this field. The earliest record we have of the project dates from March 1852, when the BAAS Council asked the Kew Committee to undertake balloon observations. The Kew Committee's main objects were to determine the temperature and humidity of the atmosphere at different heights and to see whether the chemical composition of the air at high altitudes differed from that at sea level. Arrangements were quickly made to hire Green and his *Nassau* balloon. A special set of instruments was made by London instrument maker Patrick Adie and set up by Welsh for use in the balloon. Prior to each ascent, circulars were sent out to volunteer meteorological observers across the south of England, asking them to make detailed observations on an hourly basis for the duration of each flight. The ascents were funded by the Royal Society through a £261 grant from its Wollaston Donation-fund, a private fund bequeathed by former president William Hyde Wollaston.[137]

A total of four ascents were made, between August and November 1852. Welsh accompanied Green on the balloon on each occasion, and on the first two flights Richard Nicklin, an assistant at Kew, helped with the instrument readings. The highest altitude, 22,930 feet, was attained on the fourth ascent; on this occasion, the effects of high altitude were enough to cause difficulty in breathing and tiredness after any exertion. All four ascents were a scientific success, however. In a substantial paper published in the *Philosophical Transactions of the Royal Society*, Welsh reported that the air temperature steadily declined with altitude except for a hiatus where for 2,000 feet the temperature remained static, the height of this zone varying between 4,000 and 8,000 feet depending on the weather. An analysis of air samples by the chemist William Allen Miller (a member of the Kew Committee who had helped in the early stages of the thermometer standardization work) demonstrated that the composition of the air at the highest altitudes reached was the same as that at sea level.[138]

The ascents gained some publicity outside scientific circles, as in a short article in the *Illustrated London News* with an illustration showing Welsh, Green, Nicklin, and Adie just before the first flight.[139] Yet they generated nothing like the public excitement that arose when BAAS revived meteorological balloon flights in the early 1860s, this time under a special "Balloon Committee" appointed at the 1858 annual meeting.[140] By the time the flights recommenced, Welsh was dead and the meteorologist on board was Greenwich Observatory's James Glaisher, who had taken a great interest in the 1852 flights as one of the ground observers. When the opportunity arose again, Glaisher volunteered his services.[141] Glaisher's ascents became a symbol of Victorian scientific exploration in a way that Welsh's pioneering work never did. This was no doubt partly because of the September 1862 ascent, in which Glaisher and his aeronaut Henry Coxwell had a narrow escape with their lives: the balloon went out of control and ascended to a record-breaking height of approximately 37,000 feet, causing Glaisher to temporarily pass out before Coxwell managed to stop the balloon ascending. The incident led to Glaisher and Coxwell being lionized as brave, heroic explorers living up to the best Victorian values.[142]

But the main reason why Welsh's work remained less well known was more likely due to a difference in temperament between the two men. By the early 1860s, Glaisher was already a well-known public figure: since the mid-1840s, he had been writing for various newspapers about the Greenwich meteorological observations, and he frequently gave public lectures on meteorology and his own scientific exploits, occasionally to the annoyance of his superior, George Airy. Some of his scientific contemporaries frowned upon him as a self-publicist: in 1865, the botanist Joseph Hooker described him as an example of "those cattle, who live by self-glorification." Welsh, by contrast, published only in scientific journals, and his Royal Society obituary noted his diffident personality.[143] Welsh was happy to work quietly in the disciplined environment of a Victorian observatory. Nevertheless, the work of John Welsh and the Kew Committee played an important role in Glaisher's flights in that the set of instruments used by Welsh in 1852 was later requisitioned by the BAAS Balloon Committee for Glaisher's ascents.[144]

As the narrative of the balloon flights suggests, even the dynamic John Welsh could not have done all the many types of work carried out at Kew in the 1850s—magnetic and meteorological observations, standardization, instrument development, balloon experiments, and more—on his own. From 1850, at least one assistant was employed at the observatory. Richard Nicklin, who accompanied Welsh on the first two balloon flights, was working at Kew by July 1850 at the latest and left in early 1852. He helped John Welsh with reading instruments and photographing the sun.[145] Robert Beckley started work at Kew in November 1853 on the recommendation of Warren De La Rue. Beckley was much more than an assistant: he was actively involved in instrument design and construction. In 1856, he built a cup anemometer to a design by Thomas Romney Robinson, together with a device that recorded its readings. His £91 annual salary put him at the low end of the professional classes at the time and was not far below the £100 earned by Welsh when he joined the observatory.[146] He would remain at Kew until 1871.

From the mid-1850s, the Kew Committee began employing additional assistants who were more technically qualified. March 1855 saw the arrival of Dr. Hermann Halleur on William Sykes's recommendation. He originally trained as a medical doctor before serving as director of the Royal Technical School in the German city of Bochum. At Kew,

Halleur assisted with the standardization work, but he left in September 1855. The following year, he became professor of natural science at Calcutta University for a salary of £840, a figure that must have been galling for John Welsh.[147] Halleur was replaced in March 1856 by Balfour Stewart. Like Welsh, Stewart was one of James Forbes's Edinburgh graduates in natural philosophy. He was employed by the Kew Committee as an "assistant observer," suggesting an increasingly formal structure to the observatory's personnel. Stewart also remained at Kew for just a few months. In October, he returned to Edinburgh to work as an assistant to Forbes—though not before he had invented a thermometer for measuring the sum total of temperature fluctuations, which he described in a paper for the Royal Society.[148] Stewart was succeeded in October 1856 by Charles Chambers, who was recommended by the Council of the Society of Arts after Gassiot had asked the chairman of that society's council to recommend someone suitable to succeed Stewart at Kew.[149] In January 1858, fifteen-year-old George Whipple began work at Kew; initially, he was employed "to assist in the general work of the Observatory," establishing a precedent of employing boys as junior assistants in the observatory.[150]

Welsh himself was still earning only £200 per annum at the time of his early death in May 1859. This income was modest, even when compared with the £240 paid to the Astronomer Royal's deputy at Greenwich.[151] We have no record of Welsh ever complaining to his superiors on the Kew Committee about his situation. However, when Balfour Stewart gave notice of his resignation, Welsh confided in a letter to Forbes that Stewart had "put up very good naturedly with the inconveniences of this awkward place and with the uninteresting sort of work we have so often to do."[152] Yet although Welsh had to do much "uninteresting" work for relatively little, in 1857, when he was still only 32 years old, he was recognized by being elected a Fellow of the Royal Society—ten years after the 1847 reforms that had restricted the annual intake of new Fellows to fifteen, to be elected on scientific eminence alone. The citation mentioned his balloon ascents, his work on verification of barometers, and that he was "eminent as a meteorologist."[153] That the observatory was directed by an FRS from 1857 only increased its scientific prestige.

CONCLUSION

"I should like you to see Kew. It is a nice place now."[154] So Sabine concluded a letter to John Herschel in April 1857. Certainly, much had changed at Kew Observatory since 1845. Not only had its budget hugely increased, both in terms of its annual grant from BAAS and the income it was able to attract from other sources, but by the end of the 1850s, the observatory sciences practiced at Kew had multiplied to encompass a full range of cutting-edge geomagnetic and meteorological research, in addition to a large-scale program of instrument standardization that brought in significant extra money. Whereas in 1845 the only paid worker at Kew was an army sergeant on £54 per year, by 1859 the observatory had five paid staff, resulting in a total salary bill of nearly £500.[155] By the late 1850s, too, the observatory's fame and prestige had increased enormously: its name was attached to a well-known marine barometer, and its work for expeditions and overseas observatories had given it international prominence.

The introduction of the Royal Society government grant in 1849 was less of a dramatic change for Kew Observatory than it might seem. The rescue of the observatory from closure at the end of the 1840s owed more to the astuteness of Sabine and Gassiot, who together introduced a businesslike regime at Kew. In particular, they likely saw the potential of standardization to become self-funding and, even more importantly, that it could be an essential service to government and instrument makers, thus fatally weakening the case for closing Kew. The real significance of the government grant in saving Kew was that it encouraged Sabine and Gassiot to introduce the standardization program in the early 1850s and buy a dividing engine in anticipation of grant money, ensuring that the work of testing instruments for the Admiralty, the Board of Trade, and the London instrument makers went to Kew and not to Greenwich. It is also possible that John Herschel deferred closing the observatory in anticipation of the government grant before it was officially announced—though he certainly did not believe that it should ever be used to support institutions like Kew on an ongoing basis.

From the experience at Kew, then, the government grant was not a change from laissez-faire. If anything, it enhanced the laissez-faire environment in which the Kew Committee had to work in that it gave the committee one-off sums of money that it could use to start up the standardization service, which operated on a commercial basis. In any case, only a small part of the observatory's total income before 1859 came from the government grant. Its largest source of income was still the annual BAAS grant. Even one-off projects—from Ronalds's pioneering experiments in self-recording magnetic and meteorological apparatus to the spectacular balloon flights in 1852—were as often funded from private sources as they were by government. Kew Observatory in 1859 was still an independent institution controlled by Morrell and Thackray's "gentlemen of science."

Yet Kew Observatory in the late 1850s was more than just "a nice place." It was also an important place in the landscape of the physical sciences. Sabine and Gassiot, aided by the technical ingenuity of Francis Ronalds, John Welsh, and their assistants, had established at Kew a world-class magnetic and meteorological observatory with self-recording instruments like those at Greenwich. Even when he was president of BAAS, Airy was unable to stop the Kew magnetic and meteorological observations. When the British government sponsored a display of scientific instruments at the 1855 Paris exposition (Exposition Universelle), the display was dominated by instruments from Kew and not from Greenwich. The Paris exposition was an instance of how the work of Kew was becoming internationally known. During the 1850s, Welsh and his colleagues built and tested instruments for foreign observatories, and they provided training in their use to expeditionary parties. A number of distinguished overseas visitors inspected the instruments and observational work at Kew, such as Berlin Meteorological Institute director Heinrich Wilhelm Dove, who had become famous for his global temperature maps.[156] Foreign institutions, as well as British government departments, were turning to Kew for standardized instruments. By 1859, Kew was more than an imperial capital for magnetic and meteorological instruments. It was now a global center for instrument standardization as well as research in geomagnetism and meteorology.

3

"Solar Spot Mania," "Cosmical Physics," and Meteorology, 1852–1870

I believe I have been writing a great deal of nonsense—but I must confess that a solar spot mania has fairly seized me!

JOHN WELSH TO EDWARD SABINE,
23 APRIL 1852

There is a class of observations which may be called Cosmo-Physical observations, of immense importance at the present moment; and there is, moreover, a sort of preparatory scientific conviction gradually arising that we are on the eve of some grand generalization, which may coordinate many things which seem at present strangely diverse and unconnected. . . . Among these phenomena we have the physics of sunspots, magnetic and electric disturbances, and meteorological phenomena generally.

(ANON.), *ATHENAEUM*,
3 OCTOBER 1868

BY 1859, KEW OBSERVATORY HAD BECOME NATIONALLY, INDEED INTERnationally, famous as a center for instrument standardization as well as for geomagnetic and meteorological observations. Yet the 1850s also saw the beginnings of the work for which, at least among historians of science, the observatory is now best remembered: the first systematic program to photograph the sun and its then-mysterious, ever-changing dark spots. According to the standard histories, this project was started with the word of the great Sir John Herschel; then, the Kew solar photographic telescope—the "photoheliograph"—was designed by Warren De La Rue and, under the direction of Balfour Stewart, John Welsh's successor, Kew began pushing the boundaries of the emerging discipline of astrophysics.[1] The results of the Kew photographic program were used by Stewart to derive controversial theories relating the formation of

85

sunspots to planetary alignments, and they inspired studies into possible relations between sunspots and weather that assumed considerable importance in late nineteenth-century imperial governance. The results remain of interest to astronomers today working on models of the solar cycle.[2]

This chapter asks how and why solar astronomy came to Kew in the 1850s and how the photoheliograph was used in practice. It also addresses the subject of the second epigraph at the start of this chapter: What was the true nature of the complex relationship between solar, magnetic, and meteorological research at Kew under the directorship of Balfour Stewart from 1859 to 1870? Related to this is the question of Stewart's own tortuous connections with Edward Sabine, the Kew Committee, and the Meteorological Department of the Board of Trade, later renamed the Meteorological Office. The chapter argues that while the solar photography program established at Kew was of great importance in the history of observational solar physics, it was never really central to the regime directed by Sabine and others on the Kew Committee. Moreover, that regime would not allow Balfour Stewart to develop his own theoretical researches using the solar data obtained at Kew. Instead, Stewart was forced to comply with Sabine's intensive regime of data gathering and, later, a similarly empirical program of work for the Meteorological Office.

The first part briefly sets the narrative in the historical context of what had been discovered about sunspots and their supposed influences on Earth by the mid-nineteenth century. The next section argues that John Herschel's well-publicized rallying cries for systematic solar observation were only part of a much more complex story of how solar photography started at Kew. The third section discusses how the photoheliograph was used to make its most famous observations, the photographs of the 1860 total solar eclipse. The expedition to observe this eclipse was largely directed and financed by a self-funded man of science, Warren De La Rue, and so it was a private enterprise, in keeping with the prevailing laissez-faire economic attitudes. The final two sections argue that Stewart's difficult relations with Sabine were due not only to the latter's impatience with Stewart's theory-driven usage of the photoheliograph results but also to the reorganization of the Board of Trade's Meteorological Department after Robert FitzRoy's suicide in 1865. The reorganization would lead to Kew having a much more central role in the Meteorological Department and would greatly change the work regime at the observatory. Yet despite its increased involvement with a government depart-ment, Kew would remain a largely privately funded institution until 1870.

SUNSPOTS AND SUN-EARTH CONNECTIONS

Sunspots were studied seriously by Galileo and others in the early seven-teenth century with the newly invented telescope because the presence of dark spots on the solar surface seemed to contradict the Aristotelian no-tion that the sun was pure and incorruptible. Interest in the spots there-after waned—perhaps in part because in the mid-seventeenth and early eighteenth centuries there were very few spots on the sun, a period later termed the "Maunder Minimum"—and does not seem to have revived until well into the eighteenth century. From the 1770s onward, William Herschel began observing the sun closely as part of his new approach to astronomy that emphasized the "natural history" of the heavens. Her-schel believed that the sun's luminous surface was a veneer covering a dark, solid body beneath and that the sunspots were small openings in the brilliant surface. By the mid-1790s, he was also thinking seriously about the sun's effect on Earth's climate and how long the sun would continue to shine with the same "lustre." These speculations became important to Herschel after his discovery in 1800 of invisible heat rays—infrared radiation—from the sun. He believed that sunspots enhanced the sun's total heat output and that more sunspots would result in warmer weather on Earth. Herschel went back over historical sunspot observa-tions and claimed to have detected a correlation between the number of sunspots and the price of wheat: higher sunspot activity, according to Herschel, had coincided with periods of cheaper food and therefore good harvests and warmer weather. Herschel's theories met with consid-erable controversy, but they set a precedent for speculation as to whether solar activity affected terrestrial weather.[3]

Neither Herschel nor any other early observer detected any regularity in these fluctuations in sunspot activity. In the 1820s, German apoth-ecary and astronomer Heinrich Schwabe began making daily drawings of the sun's disc in an attempt to detect a proposed planet close to the sun that some astronomers thought might be the cause of mysterious perturbations in the orbit of Mercury. He found no planet, but in 1843 he claimed that the number of spots was rising and falling in a regular, ten-year cycle. He first published these findings in *Astronomische Nachrichten*, a respected journal, but nevertheless his work does not seem to have been read in the English-speaking world at the time. The discovery was, however, noted by Alexander von Humboldt, who had been influential in stimulating interest in terrestrial magnetism in the 1820s and 1830s. Humboldt wrote about Schwabe's sunspot cycle in *Cosmos*, the great summation of Humboldt's life's work in the Earth sciences. When *Cosmos* was translated into English in 1852, the discovery would have dramatic consequences for the history of Kew Observatory.[4]

John Herschel shared many of his father's beliefs as to the nature of the sun—such as it being a solid body underneath a luminous envelope—and while at the Cape of Good Hope he was eager to investigate its properties further. He invented a device that he called an "actinometer" to measure variations in the amount of heat received by Earth from the sun and by January 1837 had begun a regular program of sunspot drawings. He used a small refracting telescope to project the sun's image onto a white screen for safe viewing.[5] By March, Herschel had obtained an interesting result: he noticed that the spots seemed to traverse the sun's disc in a pair of parallel bands on either side of the solar equator, which he likened to Earth's trade winds. He was unable to find any correlation between "the more or less spotty state of the surface of the Sun" and the radiation readings of his actinometer, but he did suggest as worth investigating whether episodes of high sunspot activity coincided with increases in the frequency of the aurora, as "both phaenomena are now in full vigour and both have had a long period of repose."[6]

In his 1847 *Results of Astronomical Observations* (here abbreviated to *Cape Results*), Herschel speculated carefully on the nature and cause of sunspots, suggesting that the sun's luminous outer envelope might be a fluid that flowed between the sun's poles and equator in a manner analogous to the global circulation of Earth's atmosphere. The currents approaching the equator might, according to Herschel, displace the luminous matter beneath, exposing patches of the dark solar interior, which we see as sunspots, and the rotation of the sun would deflect these currents longitudinally, causing the spots to appear in "trade wind" bands. Herschel urged those who observed the sun to make their drawings available so that the state of the sun on any given day could be determined. He emphasized that "a systematic and continuous series of observations of the solar spots cannot be too strongly insisted on" and that observers should pool their results "to secure an *unbroken* history" of the sun's aspect. He also suggested that the new technology of photography might be used in this collaborative solar work: "And now that clock-movements have been applied to our equatorials, and that photographic delineation can supply, in the utmost perfection, the talent of the draftsman, it were much to be wished that the subject were seriously taken up as part of the regular business of observatories. An interchange of copies might perhaps take place, without recourse to the engraver, by the aid of *the Kalotype* process of Mr. Talbot, to any moderate and useful extent."[7]

In the same year as the *Cape Results* were published, Herschel proposed in the Royal Astronomical Society's *Monthly Notices* that the RAS begin collecting "an unbroken series of such [solar] drawings" made by astronomers around the world in order to gain "a knowledge of the laws which govern these mysterious phenomena, and the periods, if any, which they observe in their formation, and thence of elucidating the nature of the sun itself." Near the end of this letter, Herschel again briefly noted that "the exceeding facility with which photographic processes are executed . . . makes their execution on a given scale, and with every requisite degree of precision, easily attainable."[8]

In fact, Herschel had realized the potential of photography in accurately recording the sun's appearance as early as 1839, almost as soon as his friend (and fellow Cambridge Wrangler) William Henry Fox Talbot had communicated to him his original "photogenic drawing" or "Talbotype" process. After successfully making a photographic image of his own using a similar process, Herschel commented that it would be "a beautiful mode of making the sun represent its own spots n times a day or of mapping the moon."[9] But although he briefly mentioned photography in both the *Cape Results* and his 1847 *Monthly Notices* letter, Herschel's emphasis in his proposed collaborative effort was on drawings to be made every clear day, as he himself had pioneered at the Cape. Herschel's main aim was for continuous coverage (note the repeated use of the word "unbroken") of the ever-changing spots, regardless of the method employed. It is likely that he regarded photography, although full of exciting possibilities, as a new and untested technique that not every potential solar observer had access to. Having worked closely with Fox Talbot on the development of the photographic process in the years immediately after 1839, Herschel had no illusions as to the technical challenges posed by this new medium.[10]

Herschel's 1847 proposal for collaborative solar observation, although published in what had become one of the world's premier journals dedicated to astronomy, was not immediately taken up. By the late 1840s, Herschel had largely retired from systematic astronomical observation, and from 1850 he would be immersed in his highly stressful job as master of the Royal Mint. Only in 1854 would he repeat his call for a systematic program to observe the sunspots. This, it has hitherto been assumed, initiated the setting up of a solar photographic telescope at Kew Observatory. Yet the story of how solar photography came to Kew involved much more than the straightforward urging of John Herschel.

"SOLAR SPOT MANIA": THE ORIGINS OF SOLAR PHOTOGRAPHY AT KEW

In the spring of 1852, Sabine obtained an important result from the large volume of data generated by the Magnetic Crusade. He found that the frequency and intensity of magnetic disturbances and the mean monthly range of magnetic variations rose and fell in a ten-year cycle, with a minimum in 1843 and a maximum in 1848. A similar period had been derived slightly earlier by Munich astronomer Johann Lamont from continental European observations, but in March 1852 Sabine noticed something else. His wife Elisabeth had recently translated Humboldt's *Cosmos* from the German, including the reference to Schwabe's discovery of a ten-year sunspot cycle. Sabine immediately saw that Schwabe's cycle coincided with the magnetic cycle that he had just discovered. He wrote excitedly to Herschel about his finding. Herschel, busy at the Royal Mint, may not have replied, for almost a month later Sabine wrote to him again, urging him "to look at the remarkable coincidence" between the sunspot and magnetic cycles, which Sabine thought was "much too remarkable & consistent . . . to be passed as a mere accident."[11] Herschel did, though, reply to Michael Faraday in November 1852, in response to Faraday forwarding him a letter from Swiss astronomer Rudolf Wolf, who had used a historical analysis of solar observations going back to the eighteenth century to refine Schwabe's period to 11.1 years. From his response, we know that Herschel acknowledged the importance of Sabine's discovery of the correlation between the sunspot and magnetic cycles: "If all this be not premature we stand on the verge of a vast cosmical discovery such as nothing hitherto imagined can compare with." He saw the discovery as a vindication of his view that a connection existed between sunspots and the aurora and suggested that electric currents in space were "auroralized" by the sun's upper atmosphere and that the "red clouds" seen during a solar eclipse (now known as solar prominences) might be "reposing auroral masses."[12]

Secondary sources suggest that Herschel was the central driving force in initiating the Kew sunspot photography program by his writing to the Kew Committee with a plea for a continuous photographic record of the sun, whereupon the Kew Committee followed his word.[13] Yet they do not explain why he only wrote to the committee in the spring of 1854, two years after Sabine's discovery of the sunspot-terrestrial magnetism correlation, nor why he recommended that Kew and not Greenwich or some private observatory take on the work. In fact, Sabine also sent his original 1852 paper on the sunspot-magnetism connection to John Welsh, the superintendent at Kew. On 23 April Welsh replied enthusiastically, agreeing with Sabine about the coincidence between the sunspot and magnetic disturbance periods. Welsh suggested that if we suppose "that the Sun is a magnet—an electro magnet—(a great electric light perhaps) it will undoubtedly have some influence upon the magnetic condition of the planet"—and that any irregularities in the sun's magnetic field would show up in Earth's. He suggested, as an initial experiment, that Schwabe's solar observations could be compared with past records of magnetic disturbances to see if outbreaks of sunspots coincided with these disturbances. Then, if a further regular series of solar data were needed, "the best way by far would be to take photographic pictures of the Sun every day at a few stations where clouds are least plentiful. These pictures are very easily obtained and require no apparatus beyond a good telescope of perhaps 10 or 12 feet focus or an object glass alone mounted on any rough thing which could be directed to the sun." Welsh went on to remark that the previous summer he had taken some images "with a very wretched little reflector which quite convinced me of the practicability of the operation." He apologized for writing what may have been nonsense, but confessed "that a solar spot mania has fairly seized me!"[14]

Although Welsh was not the first to advocate photography as a means of recording sunspots—as described previously, Herschel did so in 1839 —his April 1852 letter to Sabine is the first recorded suggestion that photography be used as the *main* method in a regular, multiobservatory photographic patrol of solar activity. He was also the first to suggest that it could be done with a small and simple, and so inexpensive, telescope, which would have been an attractive proposition to the ever budget-conscious Kew Committee. More importantly, it was Welsh who was the first to argue for a photographic sunspot observing effort in response to Sabine's discovery of the sunspot-geomagnetism relationship.

If Sabine replied to Welsh, we have no record of it. Nor do we have any evidence of correspondence or meetings between Welsh and Herschel at or before this time. Herschel did not make any move until two years later, when he wrote his well-known letter to the chairman of the Kew Committee, John Peter Gassiot. This was published in two prominent places: BAAS's 1854 annual report and the *Monthly Notices* of the RAS for March 1855. The *Monthly Notices* version, although concerned solely with Herschel's proposal for the sunspot observations, bore the generic title "On the Application of Photography to Astronomical Observations." This, together with its author's name and reputation, would have ensured the article a wide audience. Herschel emphasized the importance of a system of daily solar photographs (no mention of drawings now) in order to maintain "a consecutive and perfectly faithful record of the history of the Spots." Like Welsh, he sketched out how it could be done in practice using a small telescope (he suggested a 3-inch refractor), a magnified image of the sun being formed on the photographic paper or glass with an eyepiece equipped with wires that would enable the positions and sizes of the spots on the photographs to be measured. He also suggested, as did Welsh, that it should be done at multiple observatories in order to obtain continuous coverage, though he went further in thinking that these stations should be spread evenly in longitude around the globe.[15]

Although the letter suggests that Herschel independently took the initiative of urging the Kew Committee to take up the systematic study of sunspots using photography, secondary sources seem to have overlooked a claim in the committee's 1854 report just before the start of Herschel's letter to the effect that Sabine had reported Herschel suggesting to him the importance of daily solar photographs and that Herschel's letter was in response to one from Gassiot asking Herschel for a statement of his views on the matter.[16] The minutes of the Kew Committee for 15 March 1854 bear this out: having heard Sabine's report of Herschel's suggestion, the committee requested Gassiot to ask Herschel for "his views as to the importance of the object and the best mode of carrying it into effect."[17] No copy of Gassiot's letter seems to have survived in Herschel's correspondence, but in the original manuscript of his reply Herschel wrote, "I am ashamed to have allowed your letter to remain so long unanswered but I hope you will not attribute the delay to wilful negligence."[18] This apology is omitted from both published versions, perhaps to save Herschel the embarrassment of advertising his tardiness.

That the initiative seems to have come from Sabine and the Kew Committee helps to explain why solar photography went to Kew. It is also possible that Sabine originally owed the idea of a photographic program as much to Welsh as to Herschel and that he merely used the latter's imprimatur to gain support for the idea. Sabine's claim that Herschel suggested to him the importance of daily solar photographs does not prove that Herschel was the only one who made this suggestion to him. Furthermore, given that Herschel was overworked at the Royal Mint, it is even possible that Sabine had been informally lobbying Herschel for his support ever since his correspondence with Welsh in 1852 but that Herschel had never responded and so Sabine now asked Gassiot to write a formal letter on behalf of the Kew Committee.[19] The 1998 *Calendar* of Herschel's correspondence records very few letters on scientific subjects in the early 1850s; most of them are about Royal Mint business.[20] It is possible, then, that the original initiative came from Sabine and Welsh in 1852, not from Herschel in 1854.

Events moved quickly after the Kew Committee received Herschel's letter. Within just over two weeks, the committee had requested estimates of the costs of an instrument along the lines suggested by Herschel from the instrument makers Thomas Cooke of York and Andrew Ross of London. By the time of the BAAS annual meeting in September 1854, Ross had won the contract and the instrument was under construction. Meanwhile, a grant of £150 to cover the cost of construction was rapidly secured from the Royal Society's donation fund.[21] The donation fund was from private money, so although the Kew sunspot program is sometimes referred to as being the work of a public observatory,[22] the solar telescope was in fact funded from private sources.

The detailed design of the telescope was also worked out by a private individual, Warren De La Rue, a wealthy printer and stationer who had invented an innovative machine for making envelopes. He was also renowned as a chemist and a pioneer of astronomical photography. De La Rue had taken up astronomical photography seriously as recently as 1851, when a fine photograph of the moon on display at that year's Great Exhibition encouraged him to try his own hand at lunar photography. Astronomical photography was not new in the 1850s: photographs of the moon had been taken in the United States as early as 1840, and images of the sun, showing spots, had been produced in the mid-1840s using the French "daguerreotype" process, which recorded photographic images on sensitized metal plates.[23] But in the early 1850s, De La Rue was among the first to exploit a new photographic technology, the "wet collodion" process invented by Frederick Scott Archer. This new process greatly increased the possibilities for photographing celestial objects because plates coated with wet collodion were much more sensitive to light than daguerreotype plates or surfaces prepared with Talbot's processes. Whereas with the older processes even a bright object such as the moon required an exposure of many minutes or even hours, a good image of the moon could be secured after just a few seconds' exposure with collodion. De La Rue quickly brought himself up to speed with the new process and began taking outstanding pictures of the moon with his private 13-inch reflector. By the mid-1850s, he was widely acknowledged as Britain's leading expert on astronomical photography.[24] He was elected a member of the Kew Committee in March 1854, and it was to him that the committee turned for a detailed costing and design of a telescope dedicated to photographing the sun.[25]

Under De La Rue's direction, the telescope for photographing the sun, or "photoheliograph" as it became known, was built fairly quickly. The photoheliograph was a refracting telescope with an object glass 3.4 inches in diameter and 50 inches in focal length, which used eyepieces to project magnified images of the sun onto a photographic plate.[26] In October 1856, De La Rue reported to Herschel that the instrument was complete and erected in the dome at Kew Observatory formerly used for Francis Ronalds's atmospheric electricity observations.[27] Yet the telescope did not become operational until mid-March 1858, and it would not be used continuously at Kew to take daily solar photographs, as suggested by Welsh and Herschel, until five years after that date. It is likely that although Sabine discovered the correlation between solar activity and terrestrial magnetic variations and so had an interest in seeing his results vindicated by the photographic program, the observatory had more urgent priorities in these years. In 1857 and 1858 Welsh was doing double duty: in addition to superintending the observatory, he was busy making observations for a magnetic survey of Scotland being conducted by Sabine—indeed, De La Rue mentioned the survey as a likely cause of delay in starting the solar work at Kew.[28] At the same time, Welsh had the complex additional task of setting up the new self-recording magnetic instruments designed to work in tandem with the photoheliograph.

John Welsh's final illness that led to his early death in May 1859 further delayed the project by at least six months, but even after Balfour Stewart succeeded him as superintendent the following July, the photoheliograph was only being used intermittently. Some of the problems were of a technical nature: for example, the sun's intense radiation was causing stains to appear on the plates.[29] Then, for several months in 1860 the instrument was away from Kew for an expedition to Spain to photograph a total solar eclipse. A more fundamental problem, though, seems to have been a lack of available labor at the observatory to make regular use of the instrument. In November 1859, the Kew Committee reported that "occasional" solar photographs were being taken as opportunities arose, but that the work was "necessarily much retarded for the want of a Photographer."[30] The wet collodion process, though it enabled much shorter exposures than older photographic methods, had a disadvantage in that it was very labor intensive: the plates had to be prepared immediately prior to exposure and then exposed and developed while still wet. Therefore, two people were needed to produce solar images: one to take the pictures at the telescope, the other to prepare and develop the plates. But the Kew Committee did not have sufficient funds to hire an extra assistant to help with the work. On the eve of the 1860 eclipse, the committee's report lamented that "unless a special grant be obtained, the Photoheliograph will remain very little used."[31] In 1861, after the instrument was back from Spain, the Kew Committee decided that operating the photoheliograph interfered too much with the observatory's already very busy schedule and persuaded De La Rue to set up and run the instrument at his private observatory at Cranford, to the west of London.[32]

For an entire year, starting in February 1862, De La Rue used the photoheliograph at Cranford to take regular photographs of the sun's disc showing spots. By this time, the technical problems had been ironed out and the instrument was working to De La Rue's immense satisfaction: he considered that it could take images "so perfect that much light would be thrown on the physical constitution of the Sun if the instrument were worked 'with a will' for a few years."[33] The Kew Committee reported that the solar photographs at Cranford were taken by a "Mr. Reynolds," an assistant of De La Rue who had also helped with photographing the July 1860 eclipse. Reynolds was presumably the recipient of a £40 grant that the BAAS General Committee agreed to pay for a photographic assistant in September 1861.[34] Unlike the various salaried assistants at Kew, Reynolds is not itemized in the accounts included with the annual reports of the Kew Committee in the early 1860s, suggesting that the money was paid privately to De La Rue rather than via the Kew Committee.

De La Rue had no intention of keeping the photoheliograph permanently at Cranford: in September 1862 he expressed the hope that by the time it had completed a year's work there, steps would have been taken to have it working at Kew or somewhere else.[35] Solar photography at Cranford was duly terminated in February 1863, and by early May the instrument had begun work back at Kew. Soon after then, the labor shortage problem was solved by employing "a qualified assistant" to help with the photographic work.[36] This assistant is not named in the council minutes or the Kew Committee reports, but in 1866 De La Rue revealed that the photographs were being taken by a "Miss Beckly" [sic]. Elizabeth Beckley was the daughter of Robert Beckley, the mechanical assistant at Kew who, among other things, had built the anemometer mounted on top of the dome. De La Rue claimed that the photography "seems to be a work peculiarly fitting to a lady. During the day she watches for opportunities for photographing the Sun with that patience for which the sex is distinguished, and she never lets an opportunity escape her. It is extraordinary that even on very cloudy days, between gaps of cloud, when it would be imagined that it was almost impossible to get a photograph, yet there is always a record at Kew."[37]

Robert Beckley lived with his family in the observatory, with his wife acting as the building's housekeeper. It is likely that Sabine had some part in the idea of employing Elizabeth Beckley, for in 1861 he suggested "that Mr. Beckley should have an assistant in his department of working in wood and metals, and in photographic work: *to be found, if possible, by Mr Beckley himself*" (my emphasis). Significantly, Elizabeth Beckley's name does not appear on the salary list in the observatory's annual accounts, nor is there any mention after 1863 of a grant to employ a photographic assistant. Yet a diary kept at Kew in the 1860s reveals occasional payments of £5 to "Miss Beckley," suggesting that she was paid piecemeal.[38] Sabine, with his customary political adroitness, had devolved responsibility for the labor shortage problem to Beckley, who had solved it using casual labor from within his own family.

While woodwork and metalwork would doubtless not have been thought "fitting to a lady," it is not difficult to imagine a father-daughter partnership at work in photographing the sun. The daughter might have watched for precious intervals of clear sky and operated the photoheliograph while her father prepared the plates and developed them after exposure. This is not, of course, the first time that a woman had been employed as a scientific assistant. Indeed, going back to at least the seventeenth century, women sometimes did much more than assist male astronomers: Elisabetha Hevelius in the seventeenth century actively made and analyzed astronomical observations with her husband Johannes Hevelius, as did Caroline Herschel with her brother William a century later.[39] Moreover, in the early 1870s Elizabeth Beckley would also help with analyzing the results from the sunspot photographs. However, this seems to be the earliest case of a woman being employed in the day-to-day work of astronomical *photography*. Although women did not have a recognized role in scientific photography in the 1860s, they were by no means excluded from the burgeoning field of commercial photography. Some prominent portrait photographers in the mid-nineteenth century were women, and one contemporary source notes that by 1873 women accounted for one-third of all photographic assistants.[40] In 1880s colonial India, a labor shortage problem would again be solved when meteorological observations in Madras came to be reported to the government by Elizabeth Iris Pogson, daughter of Madras astronomer Norman Pogson, who had insufficient staff to cope with the extra burden of meteorological work.[41] It is plausible, then, that the unofficial employment of Miss Beckley as an unnamed assistant reflected a contemporary trend.

THE KEW PHOTOHELIOGRAPH IN PRACTICE: THE 1860 SOLAR ECLIPSE AND BEYOND

The most important single scientific result ever obtained with the Kew photoheliograph was derived from photographs of the total solar eclipse of 18 July 1860, which was visible over southern Europe. Total eclipses of the sun are rare events, visible only from narrow strips of Earth's surface. Very often, they can only be seen from remote locations. In the mid-nineteenth century, therefore, expeditions to observe eclipses were difficult and costly to organize. Interest in eclipses had increased slowly but steadily among astronomers since the early nineteenth century. Francis Baily's discovery of strange "beads" of light at the edge of the moon during an eclipse in 1836 had drawn the Astronomer Royal George Airy's attention to eclipses. Airy personally observed the total eclipses of 1842 in Italy and 1851 in Sweden. During both events, Airy and others had been particularly intrigued by the mysterious red flames, known as "prominences," seen around the edge of the moon's silhouette when the eclipse was total. It was not then known whether they were part of the sun or fires on the surface of the Moon—or even caused by Earth's atmosphere.[42]

The impetus to photograph the 1860 eclipse with the photoheliograph came from Warren De La Rue, who was one of those intrigued by the prominences. To photograph the prominences in sufficient detail, an ordinary telescope would not do because it gave too small an image scale. This led him to decide to use the Kew photoheliograph, whose high magnification, combined with the sensitive collodion plates, might render images of sufficient quality to solve the mystery of the prominences.[43] De La Rue obtained the Kew Committee's permission to use the photoheliograph, which was then insufficiently used at Kew due to the shortage of labor.[44] According to De La Rue, Airy offered, on his own initiative, to apply to the Admiralty for the use of a ship to convey the party of astronomers (including Airy) to northern Spain, where the total eclipse would be visible. Airy later confirmed that he had a successful interview with the First Lord of the Admiralty, which led to the astronomers being offered the HMS *Himalaya* for this purpose.[45]

The project to observe and photograph the 1860 solar eclipse was complex to organize. A special hut had to be built to house not only the photoheliograph but also a small darkroom immediately adjacent to the instrument so that the wet collodion plates could be prepared just before exposure and developed immediately afterward. This portable "photographic observatory" was assembled from prefabricated parts made in England. Several sets of photographic chemicals had to be taken to guard against the failure of one set in the remote observing location.[46] As Alex Soojung-Kim Pang has shown, Britain's imperial infrastructure was often important to the success of eclipse expeditions.[47] However, the 1860 observing effort was further complicated by its location outside the British Empire, which meant that the Foreign Office had to negotiate customs barriers. In addition, De La Rue's party had to rely on a network of local geographical knowledge, at the center of which was engineer Charles Vignoles, who at the time was helping to build the Spanish railway network.[48] The expedition was nevertheless an outstanding success. De La Rue and his assistants duly set up the photoheliograph and hut at the village of Rivabellosa in northern Spain, and the total phase of the eclipse was seen in a clear sky. Two photographs taken at different stages of totality successfully showed the prominences. The movement of the moon over the prominences during the course of the eclipse, plus the perfect coincidence of the positions of the prominences on both photographs, confirmed for De La Rue the solar origin of these features. Further corroboration was provided by photographs taken nearby by Angelo Secchi (director of the observatory at the Roman College) and others.[49]

The harmonious involvement of Airy with the Kew expedition might seem to sit uncomfortably with the story told so far of Airy's hostility to the (to him) upstart observatory on the other side of London from Greenwich. However, it could be argued that the 1860 eclipse expedition was not really a Kew expedition at all. Of the four men who assisted De La Rue with his eclipse photographs only one, Robert Beckley, appears on the Kew Observatory salary lists. Two of the assistants gave of their services for free and the other, Reynolds, was privately employed by De La Rue.[50] Balfour Stewart has no recorded involvement with the eclipse. Neither does Sabine, quite likely because the eclipse was not a direct part of his magnetic and meteorological work. On 7 July, the day the *Himalaya* set sail for Spain, Sabine was writing to Stewart about the need to start continuous photographic recording with Francis Ronalds's self-recording barometer—a thoroughly routine aspect of the work at Kew.[51] The expedition was partly funded by £150 from the Royal Society government grant; apart from the loan of a ship, this was the extent of the government's involvement. In any event, the total cost of the expedition amounted to £512, the balance of £362 being paid by De La Rue.[52] The 1860 eclipse expedition can therefore be seen not as the work of Kew Observatory but as an old-fashioned partnership between government and wealthy gentlemen-scientists, in which the government provided sea transport and diplomatic support but the science was funded largely by the practitioners themselves. In this respect it bore some similarity to the transit of Venus expedition of 1769, to which the wealthy Joseph Banks contributed much financial support, and to Herschel's self-funded expedition to the Cape in the 1830s, in which he carried out work for the government on matters such as education in the colony.[53]

The Kew photoheliograph did, nevertheless, have a great influence on solar astronomy throughout the rest of the nineteenth century and beyond in that it set a precedent for solar photography at other observatories. By the mid-1870s, "photoheliograph" would be the name for a number of telescopes dedicated to solar photography, of which many would be based closely on the original Kew design. The earliest of these was set up at the observatory in Wilna, Russia (now Vilnius, Lithuania). The Wilna photoheliograph may have had its origins in an 1858 visit by De La Rue to Russia, where he had learned about the upcoming 1860 eclipse. By the summer of 1862 John Henry Dallmeyer—the successor to Andrew Ross, who had built the Kew photoheliograph—was building a similar instrument for Wilna, under De La Rue's direction. The Wilna observatory's director, Georg Sabler, visited De La Rue to receive training on the Kew instrument.[54] The Wilna photoheliograph was working two years later and, with a brief interruption caused by the illness and death of Sabler and his assistant, remained in operation throughout the rest of the 1860s.[55] Airy acquired the Kew photoheliograph in 1873 in order to take daily solar photographs at Greenwich after the solar photography program at Kew was terminated. Two years later, he replaced it with a new telescope closely modeled on the Kew instrument. In the same way in which it had become prestigious for instrument standardization, geomagnetic observations, and meteorology, by the mid-1870s the name "Kew" had become associated with solar photography.

BALFOUR STEWART AND SUN-EARTH CONNECTIONS

John Welsh did not have the chance to direct the Kew sunspot program in practice. By 1858, he was suffering from serious health problems that prevented him from completing the magnetic survey of Scotland. He was also unable to attend that year's BAAS annual meeting at Leeds, writing to Sabine that he was "only fit to be handed over to the Doctor." In late November, in the hope of improving his condition, he left Kew to stay in Falmouth, Cornwall, with Samuel Fox, a friend and magnetic colleague of Sabine—a further indication of Welsh's closeness to the latter. A letter to his doctor indicated possible tuberculosis in several places as well as the lungs. He remained active through his illness, but he did not recover. He died on 11 May 1859, aged thirty-four.[56]

Balfour Stewart (figure 3.1) began work at Kew on 1 July; a month later the Kew Committee officially appointed him as John Welsh's successor. According to the Kew Committee, Stewart's experience as an assistant at Kew during 1856 made him "peculiarly fitted" to be the new superintendent.[57] He was just four years younger than Welsh and in some ways had a similar background, having been born into a middle-class family of Scottish merchants and studying natural philosophy at Edinburgh under James Forbes. He might therefore seem to be Welsh's "natural successor" at Kew,[58] but in fact the committee reported that Stewart was one of six applicants for the post. One of the other applicants was James Breen, who for eleven years had worked as an assistant to James Challis, Plumian Professor of Astronomy at Cambridge University. Breen's application had come with Challis's recommendation.[59] That it attracted such an experienced applicant—and the recommendation of Cambridge's Plumian professor—is a sign that the most senior paid position at Kew was becom-

FIGURE 3.1. Balfour Stewart FRS (1828–1887), superintendent of Kew
Observatory, 1859–1871. Photograph by Maull and Polyblank. Image courtesy
the Royal Society.

ing highly sought after. In addition, back in 1855 James Forbes had not been unreservedly enthusiastic about Balfour Stewart when Welsh had asked about hiring him as an assistant. Forbes wrote that Stewart was scientifically competent and hoped he would be hired, but that "his manner is at first a little dry." This was after Forbes had earlier written that he wanted to see more of Stewart's capabilities before committing himself to judgment.[60] However, the committee selected Stewart for the position on the grounds that Welsh had "repeatedly expressed to the Chairman his desire to have the assistance of Mr Stewart" and that all the others on the Kew staff wanted Stewart to be appointed.[61]

In fact, Stewart was different from Welsh in some important ways. Since leaving university as a very young man, Welsh had worked under the patriarchal Sir Thomas Brisbane at the Makerstoun Observatory before moving on to become Sabine's faithful subordinate at Kew. Stewart, by contrast, had spent his first ten years after university outside science altogether, in a business career that culminated with a short spell in Australia. After his short first period at Kew, he returned to Edinburgh to work as an assistant to Forbes in teaching and laboratory work. During this period he published what turned out to be his most important work, in which he showed that radiation did not emanate just from the surface of a body but worked throughout that body—rather like absorption, to which, as Stewart showed, radiation was equivalent, anticipating Kirchhoff and Bunsen's groundbreaking 1859 radiation laws and triggering a priority dispute with Kirchhoff. In this and other work, Stewart showed a strong theoretical bent as well as considerable grounding in experiment.[62] Therefore by the time he returned to Kew in 1859, Stewart was a mature man who had already established a firm reputation as a scientific researcher. In addition, he was used to working on his own initiative in a university environment that concentrated on teaching and research— very different from the highly utilitarian work such as instrument standardization that formed the backbone of the regime at Kew. He was less likely to be comfortable in the humble position that Welsh had accepted, for all the latter's originality as an experimentalist. This needs to be taken into account when we try to understand Stewart's actions as superintendent at Kew over the next decade.

Stewart's career at Kew began, almost literally, with a bang. On the morning of 1 September 1859, Richard Carrington, scion of a wealthy brewing family with the means to practice astronomy full time, was routinely observing the sun from his observatory at Redhill, Surrey, when he noticed a pair of intensely bright points of light appear in front of a large sunspot group. Over the next five minutes, these moved perceptibly across the sunspot before fading from view. At the time, Carrington was one of Britain's most respected observational astronomers, with a reputation as a meticulous cataloger of star positions as well as a solar observer. His September 1859 observation was part of a long-term program of visual sunspot recordings in the manner of John Herschel, plotting the positions of the spots on a projected image. Carrington carefully noted the start and finish times of this "singular appearance," as he called it. His observation of the brilliant points of light was confirmed by another British astronomer, Richard Hodgson, who had been observing the sun at the same time, and both later presented accounts of their observations at a meeting of the Royal Astronomical Society. One or more days after 1 September, Carrington visited Kew Observatory, presumably to see whether any photographs had been taken of the event with the photoheliograph. According to a later account by Charles Chambers, by then the chief magnetic observer at Kew, Stewart was away when Carrington called, so Chambers himself received him. As the photoheliograph was then still being used only intermittently, no pictures of the sun had been taken on 1 September. However, for several days before and afterward the self-recording magnetometers had registered wild variations, an event known as a "magnetic storm," and great displays of the aurora were seen from many parts of the world, including London. When Carrington and Chambers examined the magnetometer traces for 1 September, they immediately noticed a pronounced jump at the exact time when Carrington had seen the points of light on the sun.[63]

The phenomenon witnessed by Carrington and Hodgson,[64] as well as the magnetic disturbances in the days around it, seems to have sparked Stewart's interest in the sun and its relationship with terrestrial magnetism, as Stewart's published work and correspondence after 1859 shows a distinct turning toward this topic. In an 1861 paper published in the Royal Society's *Proceedings*, Stewart described the 1859 disturbances as recorded at Kew and proposed a possible explanation: that the longer-term disturbances lasting hours were caused by a large "primary" electric current emanating from the sun, while shorter and more sudden disturbances were due to slight variations in this current that induced secondary currents on the earth's surface and atmosphere, the latter causing aurorae.[65] Sabine, in a powerful position in 1861 as president of the Royal Society, took great interest in the Carrington phenomenon and initially showed broad agreement with many of Stewart's views on the magnetic disturbances. Sabine was happy to speculate with Stewart about the 1859 event. He himself wrote to Stewart about a "curious theory" advanced by Emmanuel Liais, chief meteorologist at Paris Observatory, which proposed that the sun's heat was being continually replenished by "aerolites" (meteors) falling into it; the friction of these bodies falling through the solar atmosphere caused electricity and hence sunspots and magnetic disturbances. Moreover, the incidence of these aerolites, and hence sunspots, was "regulated by the attractions of the planets near the Sun, and to have then nearly a decennial period, (due to the great influence of Jupiter)." In the same letter, Sabine wrote approvingly of Stewart's analysis of the 1859 event.[66] Similarly, when Stewart thought that he had found a correlation between sunspot activity in the sun's southern hemisphere and magnetic disturbances on Earth, Sabine thought his enquiry "well worthy of being followed up."[67]

More dramatic was Stewart and Sabine's agreement on a relationship between solar activity and displays of the aurora. In August 1862, Stewart revived (without acknowledgment) Herschel's 1852 assertion that the red flames seen during eclipses might be aurorae on the sun. Sabine responded enthusiastically and took the speculation further, suggesting that the solar "aurorae" triggered aurorae on Earth, and wondered whether "all the planets participate in such appearances, though we may never attain to their observation." Stewart expressed himself "delighted" at Sabine's agreement and suggested a variety of observational evidence in favor of the red flames on the sun being aurorae, including their red color, their possibly changeable appearance, and their greatest frequency coinciding with periods of magnetic disturbance on Earth. As to Sabine's suggestion that aurorae might occur on all the planets, Stewart wondered whether "perhaps Mr De La Rue could photograph one [of the planets] during an Aurora and ascertain this."[68] Stewart publicized his views on the red solar "protuberances" in *Philosophical Magazine*, a scientific periodical which, as Crosbie Smith has noted, had a large readership and was receptive to articles with a speculative element. Stewart gave full acknowledgment to Sabine for the idea that the sun could trigger aurorae on other planets. Photographic technology was not then capable of photographing aurorae on other planets, but in 1997 the Hubble Space Telescope photographed aurorae around the poles of Saturn, believed to be caused by particles from the sun—over a century after this initial prediction by Sabine and Stewart (figure 3.2).[69]

FIGURE 3.2. (*above*) Part of a letter from Balfour Stewart to Edward Sabine, 1 September 1862 (RS.Sa.1481), showing Stewart's prediction of the possibility of photographing aurorae on other planets. Photograph by Lee Macdonald; used with the permission of the Royal Society. (*at left*) The planet Saturn, imaged by the Hubble Space Telescope in September 1997, showing aurorae at both poles, caused by emissions from the sun—as suggested by Sabine and Stewart 135 years earlier. Image courtesy NASA.

It was partly for his work on relations between aurorae, magnetic disturbances, and terrestrial electricity that Stewart was elected a Fellow of the Royal Society (FRS) in June 1862, thus putting him in the same scientific rank as John Welsh, his deceased predecessor at Kew.[70] Yet his election certificate and his 1 September 1862 letter to Sabine both indicate another preoccupation of Stewart's: fundamental physics. Stewart's 1850s work on radiation formed part of the citation for him being elected FRS. Now, in his reply to Sabine, Stewart envisioned the electric current emitted by the sun as being in two components, one moving toward and the other away from Earth, and that the total magnetic action experienced on Earth was the difference between the two components. He suggested that sunspots could be breaks in one of these currents and that such a break represented a small change in the value of one of these components that could, in his words, "produce very powerful effects."[71] This idea that small changes in the sun's output could produce powerful effects elsewhere in the universe would shortly become important in Stewart's scientific writings.

Stewart soon went much further in his explorations into the causes and effects of sunspots. Beginning in 1864, he collaborated on a series of papers for the Royal Society that carefully described the changing positions of the spots and levels of sunspot activity, the latter being estimated by measuring the total area of the sun's surface covered by sunspots as shown on the photoheliograph images. Stewart's coinvestigators were De La Rue, who designed a special machine for measuring the sunspot areas, and Benjamin Loewy, an assistant responsible for reducing the measurements to a publishable form. Stewart claims to have taken on Loewy in early 1864 at a salary of £60 per annum, yet he does not appear on any of the Kew salary lists published in the BAAS reports. It is likely that his salary was paid by the wealthy De La Rue, especially as some of Loewy's correspondence with Stewart bears the letterhead of De La Rue's Cranford observatory. Little is known of Loewy's life, though we do know that before coming to Kew he worked as an assistant at the Flagstaff Observatory in Melbourne, Australia, a magnetic and meteorological observatory whose regime seems to have been similar to that at Kew in its early years. According to Arthur Schuster, Loewy had come to Kew on the recommendation of the Flagstaff Observatory's director, Georg Balthasar von Neumayer.[72]

From the early 1860s, Stewart also began using the Kew sunspot results in support of his theoretical researches. In April 1864, he read a paper to the Royal Society of Edinburgh in which he claimed to have found, using data from photoheliograph images, a correlation between outbreaks of sunspots at certain longitudes of the sun's surface and the drawing away of planets from above those longitudes. Stewart suggested that spot production was suppressed when a planet approached the sun due to "the preferential radiation" of the smaller body toward the larger one. This radiation, according to Stewart, was caused by the smaller body's motion through the ether—the invisible, all-pervading medium which by the mid-nineteenth century had become important in theories of light and heat propagation thanks to the widespread acceptance of the wave theory of light and the application of analytical mathematics to the science of optics. At the vanguard of this mathematical ether physics were Stewart's fellow Scottish natural philosophers James Clerk Maxwell and Peter Guthrie Tait. Stewart gave Tait partial credit for the idea of bodies radiating due to their motion. In his paper, he likened the radiative effects of moving bodies in space to those of atoms.[73] Three months later, in a letter to Sabine about magnetic disturbances, he expressed his belief that an understanding of motions and their effects on the interplanetary scale might in fact help in understanding motions at a molecular level. Stewart thought that if the motions of molecules were analogous to those of the planets, then "it is of the utmost importance to study those cosmical forces [illegible word] on a large scale which we have the opportunity of doing in order perhaps to arrive at the nature of those which act on so small a scale that we cannot directly investigate them."[74] This is the earliest recorded instance in which Stewart used the word "cosmical" when referring to fundamental physical theory. In the years to come, he would return to this "cosmical" theme when attempting to explain relationships between the sun, geomagnetism, and meteorology, notably in his inaugural lecture when he took up the professorship of natural philosophy at Owens College, Manchester, in 1870.[75]

For Stewart, the ether was crucial to understanding these "cosmical forces." Stewart and Tait believed that moving bodies dissipated their energy via friction with the ether. In December 1863, Stewart informed Sabine of an experiment that appeared to show that a rapidly rotating disc had a slightly higher temperature than a stationary one, due to its motion "through etherial [sic] space." He now proposed using part of a £50 Royal Society donation fund grant to construct an apparatus to test the heating of a rotating disc in a vacuum, thus eliminating friction with the air as a source of heating.[76] The results were published in two papers in *Proceedings of the Royal Society* in 1865; again the rotating disc produced a small yet detectable heating effect.[77] In a letter to Sabine—though not in either paper—Stewart asserted, "If we thoroughly prove these results they will I fancy be connected with celestial appearances and sun spots . . . ," again making a link between the sunspot measurements and Stewart's wider physical theory. He also asked Sabine whether he might be given an additional grant of £25 in order to complete the investigation.[78] Two months later, Stewart was again asking for money ("I do not think that more than £100 would be necessary") for another experiment to test the idea that a small body radiated more heat (via the ether) when close to a large body than when farther away.[79]

Crosbie Smith and Graeme Gooday have argued that the driving influence behind all these ideas were the strong religious convictions of Stewart, Tait, and other Scottish natural philosophers in what Smith has termed the "North British" group. Members of this group saw a need to defend a Christian cosmology against natural philosophers of a materialist persuasion, most notably John Tyndall, who exploited the now fashionable principle of the conservation of energy to argue for a purely deterministic universe. The North British group believed, in accordance with the scriptures, that the universe had a finite lifetime so that all the energy contained therein would eventually have to be dissipated. The ether offered a medium through which energy could be dissipated without violating the principle of the conservation of energy. Stewart's best-known defense of his Christian cosmology was his 1875 popular book *The Unseen Universe*, coauthored with Tait, which became a widely read statement of the case for the compatibility of science and religion.[80] Yet even in the 1860s, while he was still superintendent at Kew, Stewart was arguing for the agency of God in his cosmology. In the second of two articles for the semipopular literary and scientific *Macmillan's Magazine* published in July and August 1868, Stewart argued for a "principle of delicacy" in which the sun had an extremely sensitive molecular structure that could easily be affected by the motions of Venus and Jupiter, thus causing sunspots to form. He likened this principle to a similar one that he believed existed in human beings: human thought, he believed, required a minuscule input of energy yet could cause huge change in the same way that the tiny amount of energy involved in pulling a trigger could have a lethal effect. From this he inferred that a "Supreme Intelligence" might have a massive influence by exerting tiny amounts of energy through the "delicate" universe.[81] Something akin to this "principle of delicacy" can be discerned in Stewart's 1 September 1862 letter to Sabine discussed previously, in which he suggested that a small change in the sun's electrical output could "produce very powerful effects."

Stewart coauthored both *Macmillan's Magazine* articles with Norman Lockyer, an astronomer and science writer who would soon shoot to fame for codiscovering (with French astronomer Pierre Janssen) that the sun's "red flames" could be studied outside an eclipse by means of the spectroscope and also for the discovery of a new chemical element in the solar spectrum, which he named "helium."[82] By 1868, Lockyer was a close friend of Stewart's, and they would remain collaborators long after Lockyer founded the science journal *Nature* the following year. While it is reasonable to suppose that these two came to know each other at meetings of the Royal Astronomical Society in the early 1860s,[83] archival evidence suggests that Lockyer also visited Kew in this period. On 6 January 1863, Sabine informed Stewart that he "had an application for a Mr. Lockyer of Wimbledon to be allowed to see the Observatory. I do not know what may be the amount of his instrumental knowledge or interest." Stewart replied that he would be happy to show Lockyer around.[84] Sabine's reference to "a Mr. Lockyer" demonstrates how little known Lockyer was in the early 1860s, when he was working full time at the War Office and his astronomical activities were often restricted to evening observations from his garden in Wimbledon. If Lockyer visited Stewart at Kew sometime in January 1863, the two of them would have had the opportunity to exchange ideas in private, away from the hubbub of London scientific meetings, at precisely the time Stewart was developing his ideas on the aurora, "delicacy," and "cosmical" forces.

Exactly what Sabine thought about these theories of sunspots and ether is not known, but from the defensive tone of some of Stewart's replies we can tell that Sabine had at least some reservations. For example, his theory of planetary influences on sunspot formation must have been too much for Sabine, for in January 1865 Stewart wrote, regarding a paper on this subject that he had sent to the Royal Society, "I do not think that sun spots are the work of venus [*sic*]"; he thought merely that Venus had "the effect of *regulating* the phenomena the *predisposing* cause of which is to be looked for elsewhere."[85] His 27 April proposal to spend £100 on an experiment to test the radiation of heat from a small body at varying distances from a large one could not have been well received either—and not only by Sabine, for it seems from Stewart's 13 May response that George Gabriel Stokes, Lucasian Professor of Mathematics at Cambridge and secretary of the Royal Society, found the idea "speculative." Stewart responded with a strident reminder to Sabine that "there is no doubt from *your own observations* that there is a connexion between the sun & the earth different from a mere gravity influence I need only allude to the connexion between sun spots & magnetic disturbances."

Stewart conceded that "we tried to push the thing on too fast" and proposed to just carry on with the rotating disc experiment.[86] Stewart was still defensive about his rotating disc experiments in December 1866: referring to a less than enthusiastic response to his latest paper on this subject read to the Royal Society the previous evening, he emphasized to Sabine that the heating effect observed in the rotating disc "is a *new result*—both [William] Thomson and Maxwell consider that a *new fact* has come to light"—strategically deploying two of the most prestigious names in 1860s physics.[87]

Gooday has argued that Stewart, as a theory-driven natural philosopher, sought to understand the big picture of meteorology, geomagnetism, and solar activity as part of his "cosmical" view of physics. To this end he insisted, against the wishes of the Kew Committee, on recording levels of atmospheric water vapor as part of an attempt to understand large-scale atmospheric dynamics, and he also unsuccessfully asked Sabine for a full ten years' worth of magnetic results with a view to understanding the relationship between the sun and terrestrial magnetism over an entire sunspot cycle. Both requests brought him into conflict with Sabine and Gassiot, whose aims took the form of "a gentlemanly 'natural history' of meteorological observations" that involved collecting and presenting large amounts of data as an end in itself, with no need for any theoretical extrapolation from the results. According to Gooday, this conflict led ultimately to Stewart's resignation from the observatory. It is certainly true that Stewart conflicted with Sabine over both requests: he was only able to complete the ten-year magnetic tabulations with a donation of £400 from Gassiot's personal fortune, and he made water vapor recordings a condition of withdrawing his resignation in 1869 (see the following). A further possible reason for personal animosity between Sabine and Stewart was that Sabine was a good friend of Tyndall, Stewart's arch-opponent in the "unseen universe" debate.[88]

However, in 1865 the situation at Kew was complicated by a personal tragedy. On 30 April Robert FitzRoy, head of the Meteorological Department of the Board of Trade, committed suicide. Many contemporaries believed that FitzRoy took his own life partly due to a highly strung temperament but also because he had become demoralized by increasing criticism from scientific colleagues of the accuracy of his weather "forecasts" (FitzRoy's own term). Within days of FitzRoy's death, the Board of Trade asked the Royal Society for advice as to how the Meteorological Department should be run in the future. Sabine, as president of the Royal Society, took the initiative at once. He joined the chorus of those criticizing the "unscientific" methods of the department under FitzRoy and as early as 15 June 1865 wrote a letter to the Board of Trade, calling for a larger number of meteorological observations to be made. Hitherto the department had only collected observations made at sea, but now, to bring Britain into line with other nations, Sabine called for observations to be made on land as well and that these land observations should be collated and published at a "central office." Most importantly, in the same letter he recommended that Kew Observatory could be used "as the central meteorological station" to standardize and supply self-recording instruments to a string of land observatories distributed in a north-south line across the British Isles and to receive the resulting observations.[89]

The Board of Trade appointed a committee, with representatives from the Admiralty and the Royal Society as well as the board, to report on the Meteorological Department's current activities and draw up recommendations for its future. The Royal Society's representative on this committee was Francis Galton, who was then best known as a meteorologist and a geographer. A leading member of the Royal Geographical Society, in 1863 he had published *Meteorographica*, a book on mapping the weather. Galton chaired the committee, and so the report that it presented to Parliament on 13 April 1866 has become known to historians as the "Galton Report."[90] It criticized the department's data-gathering methods under FitzRoy and, especially, the weather forecasts. To much controversy among sailors, members of Parliament, and the wider public, the committee recommended that both the forecasts and FitzRoy's system of storm warnings (the latter especially popular with sailors) should be suspended until such time as they could be placed on a scientific basis. The collection of data, the report stipulated, should now be supervised by "a scientific body"; alternatively, Kew Observatory could be adapted for this purpose. As in Sabine's June 1865 letter, the report recommended the establishment of a series of six meteorological observatories, each with an identical set of self-recording instruments, and that Kew become the nerve center to which data from these outstations should be sent.[91]

Galton has received credit for writing the report. In support of this assertion, Katharine Anderson has noted that he was the only member of the committee who (like his cousin, Charles Darwin) had ample independent means with which to pursue his scientific interests and so had time to draw up the report.[92] But the report's recommendations for the land meteorological observations clearly bear a suspicious sim-

ilarity to Sabine's June 1865 letter to the board, especially the idea that
Kew should become the central observatory. Galton was a good friend
of Sabine's and shared the latter's predilection for gathering large quantities of statistics. He had also been a member of the Kew Committee
since 1860 and had begun supervising the testing of sextants at Kew in
connection with his Royal Geographical Society work; he seems to have
commanded enough authority to have had some stone posts erected in
the observatory grounds in connection with this.[93] A letter from Galton
to Sabine, written just after the report was published and explaining his
reasoning behind it, is particularly revealing. Galton's suggestions for
"putting the whole of the meteorology into the hands of Kew" and that
the reconstructed Meteorological Department would act as "a *branch office
in London*" reporting to Kew would have been music to Sabine's ears.[94]

That Galton probably colluded with Sabine in the so-called "Galton
Report" is consistent with Sabine's track record of maneuvering behind
the scenes to achieve his aims. In the sense that it gave Sabine control
over British land meteorology via Kew Observatory, the report is an echo
of Sabine's earlier machinations for a meteorological and magnetic observatory independent of Greenwich. Indeed, the Galton Report might
well have been another move to keep Sabine's rival, George Airy, at bay,
for its proposals did not include any role for the magnetic and meteorological department at Greenwich. Predictably, this angered George
Airy, who began a heated correspondence with the Royal Society Council, asserting that even Kew's "very respectable position" did not justify
the Council "in absolutely setting aside all notice of the Government
Observatory." The council took no action on this letter.[95] This was not
long after a separate dispute, earlier in the 1860s, in which Airy had
challenged Sabine as to the necessity of continuing the magnetic observations at Kew while the same observations were being carried out at
Greenwich. Airy claimed that he saw it as his duty "as National Observer"
to measure and print the figures shown by the Greenwich instruments.[96]
Now, the new proposals by Sabine and Galton once again challenged
Airy's prestige as the "National Observer."

Making Kew the central meteorological observatory, with responsibility for reducing, tabulating, and publishing the results from the six
proposed outstations as well as standardizing and inspecting the instruments for these observatories, would dramatically increase its workload—
and that of its superintendent. Stewart gave his reactions to the Galton
Report's proposals in three letters to Gassiot, still chairman of the Kew

Committee. Stewart said that he would decline to remain as superintendent if Kew were to be dedicated entirely to meteorology or if his job were to be divided into two posts, one running the current work being done under BAAS and the other working for the Board of Trade. But if both branches of work were to be placed under one director, Stewart said he would be happy to remain in the position. His only reservation was about the reductions of the land observations. He asked if these could be done elsewhere, as the board was planning to do with the marine observations. But the Kew Committee does not seem to have entertained this possibility, as well they might not, given Sabine's desire for control over the observations: "The Superintendent of Kew Observatory should also be *responsible reducer of all the observations*." Stewart nevertheless agreed to put his name forward to continue in the role and to "make every possible arrangement to devote my whole time to the duties of this office." He also solemnly pledged to cease work on the experiments he was doing in collaboration with Tait (presumably those connected with ether) and also to stop spending any time on the sunspot investigations after he had finished the paper he was currently writing with De La Rue and Loewy.[97]

Stewart also suggested, apparently on his own initiative, that the Kew Committee's duties under the proposed new regime would be eased if it had the services of a secretary who would "make himself acquainted with the whole concern & who might be always at hand." Stewart then volunteered himself for this position. In October 1866 the Kew Committee and the Royal Society Council held a joint meeting at which it was agreed that the Board of Trade's system of meteorological observatories would be run by a "Superintending Scientific Committee" whose members would be unpaid but which would need "a competent paid Secretary"—five months after Stewart had volunteered his services.[98] Stewart's putting himself forward and, indeed, suggesting the creation of the secretary's position, might have been because he saw it as an opportunity to increase his salary, which was then just £200 per annum. It was with the aim in mind of increasing his annual income to £400 that in 1861 he had written to Sabine to explore the possibility of his also taking on the position of secretary to BAAS.[99] By the mid-1860s, Stewart was married with a young family, so he would have had a motive to increase his income.

Stewart's hopes of a pay increase were rewarded when in January 1867 his salary was increased to £400 on the understanding that he serve both as superintendent of Kew Observatory and as secretary to the Board of

Trade's scientific committee, by now renamed the Meteorological Committee. This new committee, chaired by Sabine, had eight members, five of whom were either on the Kew Committee or had served on it recently. Robert Henry Scott was appointed as FitzRoy's successor at the head of the Meteorological Department, very likely through his friendship with Sabine—another indication of Sabine's desire for control over the meteorological observations. By the summer of 1868, self-recording instruments had all been verified at Kew and sent to the observatories that were established at Falmouth, Stonyhurst (Lancashire), Glasgow, Aberdeen, Armagh, and also Valencia on the west coast of Ireland.[100]

THE RESIGNATION OF BALFOUR STEWART

Stewart's fears about the increased workload at Kew soon proved to be justified, especially while he was busy reducing the Kew magnetic observations even before the new meteorological observatories started sending in their results. In January 1867, his request for some extra money to pay Loewy to help with the reductions went nowhere. Stewart's further complaints six months later brought some stern admonition from Gassiot, who explained to Stewart that his requests for assistance had always been treated with "the most liberal spirit" but that he had to remember that he was now secretary of the Meteorological Committee as well as superintendent of Kew. Gassiot ended the letter with a warning that Stewart would now have little time for experiments requiring his "*personal investigation*."[101] Stewart did manage to hire a junior assistant, but Sabine's letter of 31 July, setting out how this assistant should be employed, again has the tone of a schoolteacher disciplining a wayward pupil: while waiting for the meteorological data to come in, said Sabine, the "youth" should be employed in tabulating the magnetic results, as Stewart would find the resulting experience very valuable "when ere long, you will have to state the pecuniary aid you will require as a central meteorological Observatory." In the same letter, Sabine assured Stewart that it was not the aim of BAAS or the Kew Committee "to view the ultimate purpose of the photograms [the traces from the self-recording instruments] as accomplished by their being merely put away in drawers for safe keeping."[102] In 1876, some years after leaving Kew, in an article for Lockyer's journal *Nature*, Stewart used an identical phraseology in criticizing the meteorology of the past as having been conducted by "Royal Societies and Astronomical Institutions" (most likely a dig at the Royal Society and Kew Observatory) and whose results "were reduced after a mechanical and strictly statistical method, and then put aside in a drawer."[103] That Sabine's words should rankle Stewart such that he could throw them back in his face some nine years after they were written demonstrates the strength of Stewart's feelings about the regime at Kew.

As if all this were not enough, the standardization work begun at Kew in the early 1850s continued throughout the 1860s. Until the mid-1860s, around 100 barometers and 400 thermometers were being tested and issued with certificates at Kew each year, but by 1869 the number of thermometers tested each year had increased to over 1,100. The construction of self-recording instruments for foreign observatories —such as the Coimbra Observatory in Portugal, to which instruments were sent in 1867—also remained central to the work at Kew. A request from a brewery in 1869 for a standard thermometer suggests that Kew was also becoming recognized in the commercial sector as a source of high-quality instruments.[104] Indeed, Kew had become so well known that, as John Davis has persuasively argued, French scientists began lobbying for a dedicated magnetic and meteorological observatory along the lines of Kew; by 1870 such an institution, partly modeled on Kew, had been established at Montsouris, a mile from Paris Observatory.[105] Even before the workload increased after the Meteorological Department reorganization, the modest two-story Georgian building must have been a cramped space in which to work. With the magnetic instruments in the basement, the photoheliograph and accompanying apparatus in the dome, plus the meteorological and magnetic reductions and instrument standardization taking place in the middle floors, there was no room to spare. This lack of space was recognized in the Galton Report, and money was provided for an additional outbuilding and alterations to an existing outhouse in the observatory grounds.[106]

Despite his assurances to Gassiot in 1866 that he would give up the solar work and other independent research, Stewart continued with both. In 1869 Stewart was still applying, successfully, for grant money from the Royal Society to continue his rotating disc experiments. He also continued to coauthor papers on sunspots and their possible periodicities with De La Rue and Loewy and popularized his solar results in articles such as those in *Macmillan's Magazine* in 1868. It must therefore have been very frustrating for Stewart to have to spend so much of his time on work for the Meteorological Department (known as the Meteorological Office from 1867 on), especially with the limited complement of staff at Kew. It is in this context of extreme frustration that we need to regard his and

Lockyer's *Macmillan's* articles. As Gooday points out, to explain to their readers the concept of potential energy, Stewart and Lockyer used the analogy of the upper classes in society automatically having power over the lower orders, which in Stewart's case might well have been a metaphor for the oppressive regime imposed by the wealthy and influential Sabine.[107]

A much more direct and vehement attack on the Kew regime appeared in an anonymously authored article in the *Athenaeum* on 3 October 1868. Near the beginning, the article described "a class of observations which may be called Cosmo-Physical observations, of immense importance at the present moment" and asserted that "we are on the eve of some grand generalization" that would encompass sunspots, magnetic disturbances, and meteorology and demonstrate the connections between them (quoted in full at the head of this chapter). The article praised the "excellent" techniques with which the observations were being made at Kew and likened the observatory to "the head-quarters of an invading army" that, however, was now being starved of supplies in the form of funding to reduce all the data.[108] This military metaphor is surely a reference to Sabine, by now a general of the Royal Artillery. The article's emphasis and phraseology, especially the reference to "Cosmo-Physical observations," points the finger toward Stewart and Lockyer as authors. The article's call for increased public funding for science would soon be a hallmark of Lockyer, who after 1870 became an outspoken advocate for state support of science. Stewart's authorship is also made clear by the similarity in language between this article and one by Stewart in *Nature* a year later, in which he criticized meteorology for lacking an overall theory as to the workings of Earth's atmosphere and again opted for a military metaphor when describing our current picture of meteorology. "We are like a soldier in the midst of a great battle," Stewart wrote, "who can give but a very poor and partial account of [meteorology] . . . and ignorant, as he must be, of the general plan of the whole."[109]

Another anonymous article appeared in the *Saturday Review* on 7 November. Whereas the *Athenaeum* was aimed at learned gentlemen such as Sabine, the *Saturday Review* had a much more general readership and explained the work of Kew Observatory and the Meteorological Committee in layman's terms. Yet it pointed to exactly the same problem as had the *Athenaeum* piece: the lack of funding for meteorological reductions, "without which the observations might as well not be made." True to its intended middle-class, nonscientific readership, the article likened this policy to "a man who should spend 1,000*l.* a year in supplying his household with food, and refuse the additional 100*l.* required for fuel and cooking to fit it for use."[110] In addition, the article praised the efforts of Stewart and members of the Kew Committee while hardly mentioning the Meteorological Office's governing committee, which was chaired by Sabine and oversaw the meteorological reductions at Kew.

Both articles provoked infuriated responses from Gassiot and Sabine. On 17 November, Gassiot wrote to Lockyer that "the general tone of your friends [*sic*] article [likely the *Saturday Review* piece] is so palpably to exalt Mr. Stewart and the Kew Committee to ignore Mr. Scott, Capt. Toynbee and the Meteorological Committee that the members of the Kew Committee who are also members of the Meteorological department must take some action."[111] Stewart, for his part, claimed that he had not seen the article in the *Saturday Review*, but that "some time since I read an article in the Athenaeum in which I thought Mr. Scotts [*sic*] position was much too slightly mentioned his name not at all." He went on to suggest its potential for causing "a feeling of awkwardness" between Robert Scott and himself and expressed "much regret" at any implied comparison between himself and those in charge of the Meteorological Office.[112]

Stewart had further disagreements with Sabine and the Kew Committee during 1869. In January he made a fruitless request for an assistant dedicated to reducing the results from the outlying meteorological observatories.[113] Then in April, Stewart tried to farm out the magnetic reductions to an unnamed third party outside Kew. The pressure being borne by Stewart is plainly evident in his defense of this latter move: "I do not think any one can be more desirous than myself that the magnetical part of this establishment should be well represented. In the present position of this institution and bearing in mind what we have to do for the Meteorological Committee I have seen that [the magnetic tabulations] can be best done out of the observatory."[114]

Stewart resigned as superintendent of Kew Observatory on 8 October 1869; he also resigned as secretary to the Meteorological Committee three days later. Stewart stated no reasons in his resignation letter. A memorandum written by Gassiot records that he resigned because his health would no longer stand the pressure of work.[115] More revealing are the conditions under which Stewart offered to withdraw his resignation: first, that the meteorological reductions include some important additional elements, notably the degree of vapor and mass of dry air; and second, that he be given more assistance at Kew.[116]

Yet Stewart was never given the chance to withdraw his resignation. According to Gassiot—and also to Arthur Schuster, writing candidly some six decades after the event—as soon as Gassiot mentioned Stewart's possible resignation to Sabine, the two of them discussed the appointment of a successor. They had in mind Charles Chambers, who had started as an assistant at Kew in 1856 and had earned his spurs there by taking charge of the magnetic work during John Welsh's final illness. He had left Kew in 1863 and was now in charge of the Bombay magnetic and meteorological observatory, but he wanted to return to Europe for health reasons. Chambers was a loyal colonial observer, very much in Sabine's mold, and so would have been a natural choice of successor to Stewart. According to Gassiot, by the time Stewart had sent his letter with his conditions, Sabine had already posted a letter to Chambers's superior at Bombay, offering Chambers the position.[117] The implication is clear that Sabine and Gassiot were eager to remove Stewart from the observatory now that his initial resignation letter provided them with an excuse for doing so.

Stewart's resignation was not effective immediately. He left the date open until he had found a suitable position. This only happened the following year, when in May he applied for the vacant chair in natural philosophy at Owens College, Manchester. Schuster quotes a testimonial from Sabine in support of Stewart's application. Its tone was scathing, saying that had Stewart completed the magnetic work he was supposed to do at Kew he would have been "in a preeminent position."[118] Stewart's application was successful, however. As a parting shot, Stewart could not resist taunting Sabine. As a possible research topic for the new laboratory that was being built for him at Manchester, said Stewart, "I think of suggesting magnetism namely a set of self recording instruments the curves of which are systematically tabulated & reduced—and a set of monthly absolute observations. . . . What should you think of the value of such a series?"[119] If Sabine replied, we have no record of it.

The conflict between Stewart and Sabine cannot be attributed entirely to the reorganization of the Meteorological Department. As we have seen, friction between them over Stewart's theories on sunspot periodicities was evident by (at the latest) January 1865, before FitzRoy's suicide. Yet neither can it be put down solely to a clash between Stewart, the natural philosopher on the one hand, and, on the other, the "gentlemanly" natural historians, represented by Sabine. Making Kew the Meteorological Office's central observatory did dramatically increase Stewart's workload, especially as the magnetic reductions and instrument standardization already made for a hectic schedule. The new work, moreover, was of a very routine, utilitarian nature: gathering, collating, and publishing statistics. Stewart had done some of his most creative work under Forbes at Edinburgh, and he remained a university physicist at heart after returning to Kew, continuing with his sunspot and ether research even after the meteorological reorganization. Unfortunately for Stewart, an important part of his duties after 1867 had effectively become that of an employee of the Meteorological Office: someone paid to report statistics to the government who had little time for independent research—as Gassiot emphatically reminded him in 1867. Sabine, for his part, saw geomagnetic and meteorological research as a matter of collecting large volumes of data. He was not averse to theoretical speculation—such as on the possibility of aurorae on other planets—but he would not allow this to form the basis of research or to interfere with data gathering, which he saw as the observatory's main purpose. Sabine's effective dismissal of Stewart may be compared with his similar move against William Radcliff Birt in 1850, in the sense that neither man fitted into the position of loyal subordinate in General Sabine's troops. This was especially true now that by taking control over land observations at the Meteorological Office, Sabine had launched a "meteorological crusade" similar to his Magnetic Crusade. He had succeeded in the aim that he had had in mind since 1840: making Kew a center for both magnetism and meteorology on a national scale and independent of Greenwich.

CONCLUSION

Ever since the 1860s, Kew Observatory has been associated with the early years of solar physics—in popular as well as academic histories of science.[120] Kew was indeed gripped by what John Welsh called "solar spot mania" in the 1850s and 1860s. By 1870, the photoheliograph was being used to take more than 300 images of the sun per year.[121] It had some great successes to its credit, notably confirmation of the solar origin of the "red flames" seen during solar eclipses and a system of daily solar photography that would later be continued at Greenwich Observatory. The work increased the importance and prestige of Kew Observatory to new levels. Thanks to the "Kew photoheliograph," Kew was now a synonym for solar astronomy as well as geomagnetism, meteorology, and instrument standardization.

Nevertheless, the solar program was never really central to the work at

Kew. That a systematic program of sunspot observation came to Kew at all was not a given. The Kew photoheliograph was not a straightforward result of Herschel's rallying cries. It is clear that the instigation of the solar work at Kew also owed much to Edward Sabine and, quite possibly, John Welsh. Even after solar photography began there, however, it remained outside the observatory's central routine. The photoheliograph and the work done with it were largely funded by Royal Society grants from private sources or by Warren De La Rue, a classic example of a self-funded Victorian devotee of science. That neither Elizabeth Beckley, who helped take the solar photographs on a daily basis, nor Benjamin Loewy, who reduced the sunspot numbers and areas into a form usable for calculation, appeared on the Kew payroll reinforces the case that the photoheliograph was an example of private patronage. The priority at Kew was always Sabine's central concern with the collection of magnetic and meteorological data, in addition to the standardization work for British and overseas governments that brought in essential income.

After Balfour Stewart succeeded John Welsh as superintendent in 1859, Sabine and the Kew Committee never gave him a free hand to develop his theory-driven "cosmical physics." As both superintendent of Kew Observatory and, from 1867, secretary to the Meteorological Committee, Stewart was expected to follow Sabine's empirical style of research, amassing more and more magnetic and meteorological data. Nonetheless, he would not let go of his beloved private research. Part of Stewart's conflict with Sabine can be attributed to the differences in Stewart's background and personality from his predecessor, John Welsh. Whereas Welsh was fundamentally an experimentalist with a genius for invention and practical problem solving, Stewart was always more of a theoretician, with a natural-philosophical bent more in tune with his countrymen Tait and Maxwell than with Sabine and Gassiot, his superiors at Kew. He was also older and more experienced and so less likely to be subservient to the authoritarian Sabine. Yet it is clear also that the conflict also owed much to the changed nature of Stewart's post after 1865. The mass of data reductions for the Meteorological Department—now renamed the Meteorological Office, with Kew as its central observatory—entailed a huge increase in Stewart's workload, yet he was unable to persuade Sabine, the Kew Committee, or the Meteorological Committee to fund the extra staff that he was calling for.

By 1870, Kew Observatory had an annual income of over £1,575. Of this, over £600 came in the form of an annual grant from the Meteorological Office.[122] Where meteorology was concerned, Kew could therefore be described as a government observatory in the sense that it acted as a central observatory that supervised a network of self-recording meteorological stations across the British Isles and processed the results that they sent in. But in every other sense, Kew was *not* a government observatory. Apart from one-off sums from the Royal Society government grant, the rest of its income came from private sources and the money it made from instrument standardization. The geomagnetic observations, the solar photography, and the standardization work continued on a privately funded basis, as before. In particular, the solar photography at Kew was not the work of a public observatory, as Simon Schaffer has suggested.[123] The "public" observatory's remit was the strictly utilitarian meteorological data collection, while control of the solar research remained very much in the hands of self-funded devotees of science such as De La Rue. Even if the term *public observatory* is used in the broader sense that Kew was not a person's private property, the word *public* is misleading because it obscures the fact that Kew was mostly privately funded. With the exception of the work for the Meteorological Office, support of the sciences at Kew at the end of the 1860s continued to rely on the laissez-faire system.

4

Kew Observatory and
the Royal Society, 1869–1885

I need scarcely say that it has afforded me much pleasure, to have had it in my power,
through the Royal Society, to assist in maintaining an Establishment with which I
have, for so many years, been connected

JOHN PETER GASSIOT TO WILLIAM SHARPEY
(SECRETARY, ROYAL SOCIETY), 4 JULY 1871

a large proportion of the various thermometrical determinations made by English
physicists are dependent for their accuracy upon that of the verifications at Kew.
Many thousands of thermometers have already been verified by the apparatus about
to be described.

FRANCIS GALTON, 1877

UNTIL THE START OF THE CONFLICT BETWEEN BALFOUR STEWART AND
Edward Sabine, the 1860s were a stable and relatively prosperous period
for Kew, during which the observatory achieved worldwide fame. The
threats of closure that had hung over the observatory for most of the
first decade after the BAAS takeover in 1842 were not repeated. This
situation changed in the late 1860s when BAAS, ever short of money,
declared that it no longer wanted to keep running Kew. According to the
standard histories, two years later the Royal Society stepped in to take
over its management, thanks to a generous donation from Kew Committee chairman John Gassiot. Many sources also simply state that following this, in the early 1870s, the solar work went to Greenwich while
Kew consolidated its role as the central observatory of the Meteorological Office and continued with its work in instrument standardization.[1]

This chapter takes as its starting point the announcement in 1869 that relations between BAAS and the Kew Committee were to be reviewed. It finishes in 1885, the year that saw the publication of Robert Henry Scott's well-known general history of Kew Observatory. Scott's history has some serious limitations, but it is so widely cited that the year of its publication is a useful landmark in the history of Kew. It also conveniently divides the history of the observatory between 1869 and 1900 into two periods. This chapter's first section examines in detail how and why Kew was transferred from BAAS to the Royal Society between 1869 and 1871. I argue that the story is much more complex than has hitherto been supposed, in addition to making the case for the importance to Kew of private funding. The next two sections—on Airy's failed coup over the Kew meteorological observations and his successful one to take over the photoheliograph—show how Airy continued his battle for superiority over Edward Sabine and Kew into the 1870s. Airy's struggles with the Kew Committee over meteorology show how meteorology was coming to be governed by its own specialist institution, over which Greenwich had no control—mirroring a contemporary trend toward specialization in the observatory sciences. The case of the photoheliograph helps us to understand why Airy resigned the presidency of the Royal Society after only a year in office. The final section assesses the changed working regime and personnel at Kew from 1871 to 1885. Here I argue that during this period, a decline in income from both the Gassiot Trust and the Meteorological Office grant, in addition to the illness and death of Sabine, forced Kew to change its emphasis from geomagnetism and meteorology to testing instruments in return for fees. This, I suggest, resulted in Kew changing its character from a place of experimental investigation and data gathering into a commercially driven laboratory that served industry and government.

More fundamentally, this chapter sets Kew Observatory in the context of the 1870s debate on the relations between the British government and science. This debate centered around the government's Devonshire Commission, set up to look into the state of science education and institutions for scientific research. Some historians have interpreted the period of the Devonshire Commission as an early sign of the end of laissez-faire attitudes and greater involvement by the state in scientific research. However, a principal aim of this chapter is to show how, in the years after 1871, Kew remained an example of an institution financed in the older Victorian manner: that is, mostly from private sources, not government. This challenges Roy MacLeod's view that there was little private funding for science in the last three decades of the nineteenth century.[2]

FROM THE BRITISH ASSOCIATION TO THE ROYAL SOCIETY, 1869–1871

Having survived the various threats of closure between 1845 and 1850, throughout the 1850s and 1860s Kew Observatory was voted a substantial annual grant from BAAS, rising from £300 in 1850 to £600 in 1869. Then, at the 1869 annual meeting in Exeter, the General Committee of BAAS decided "that the existing relations between the Kew Committee and the British Association be referred to the Council to report thereon."[3] This was the first time since the 1840s that the future of Kew Observatory as a BAAS institution had been questioned, though unlike 1845, the resolution attracted no comment in the wider press such as *The Athenaeum*, so it is harder to say who might have originally moved the motion and what their motives were. It is notable, however, that the incoming BAAS president in 1869 was Thomas Huxley, who had long resented Royal Society president Sabine's alleged preference for the physical sciences over Huxley's own field of biology.[4] If Huxley could not control Sabine's activities in the Royal Society, he might have thought that becoming president of BAAS could give him more power through that organization. The £600 annually granted to Kew Observatory represented a substantial portion of the association's modest income, which Huxley might have felt could be put to better use.

On 13 November 1869, the BAAS Council appointed a special committee of thirteen members to consider the resolution and report back to the council. The committee included two biologists, Huxley and Joseph Dalton Hooker (director of the Royal Botanic Gardens, next door to Kew Observatory), as well as most members of the existing Kew Committee.[5] The special committee met on 27 November and decided that the association should continue to run Kew as before until 1872, in which year the ongoing program of simultaneously monitoring terrestrial magnetism and photographing the sun would have completed a full ten years, but that after 1872 "all connexion between them should cease."[6] The committee's report, signed by Gassiot (still chairman of the Kew Committee) and presented to the council on 11 December, confirms that the reason for the decision to review the relations between BAAS and the Kew Committee was a financial one—"whether the sum of £600, annually granted by the Association, can be reduced without impairing the efficiency of the Observatory." More particularly, the report asked whether such a reduction could be made by discontinuing some of the observatory's current work. Gassiot argued that terminating the magnetic observations would save a mere £110 a year, much of which constituted lucrative overtime for staff, who might leave if they lost this. Furthermore, the report rather bluntly stated that if the observatory were to be reduced to a mere depository for instruments, the committee could not recommend continuing the observations currently being made for the Meteorological Office. Therefore, it was not practicable to terminate either the magnetic or the meteorological work. The report concluded that the £600 annually voted on by the association could not be reduced without compromising the work of the observatory.[7]

Nothing more about Kew Observatory is mentioned in the BAAS papers until the annual report of the Kew Committee was presented at the next annual meeting, held at Liverpool in September 1870. This ratified the decision to sever the connection between BAAS and Kew effective from 1872 and made it clear that this implied considering "the dissolution of the Kew establishment."[8] In addition, the report summarized a statement by Balfour Stewart "on the past and present condition of the Observatory" and used this as evidence of how the observatory now received a large portion of its funding from sources outside BAAS, notably the Royal Society. The unwritten implication, therefore, was that there was no need for BAAS to continue supporting it. Indeed, at the same meeting, it was resolved that the president and council contact the Royal Society and the government with a view to offering the future use of the Kew buildings to the Royal Society, assuming that the Royal Society wanted them.[9]

Given that Sabine was president of the Royal Society, it is reasonable to infer that this resolution, and perhaps also the decision to terminate the connection between BAAS and Kew Observatory, amounted to a subtle maneuver by Sabine to transfer the observatory to the Royal Society and so allow him to tighten his control over it. Declaring that the observatory could not be run for less than £600 a year would have been enough to force the BAAS Council into giving it up because this annual running cost was the stated reason why BAAS's relations with Kew were being reviewed in the first place. Similarly, Sabine might have used the threat of the observatory's "dissolution" as a means of forcing the Royal Society's hand in taking it over. Many besides Sabine might well have been thinking along these lines also, for it is striking how in 1870 many prominent members of the BAAS Council—Galton, Gassiot, Sabine, William Sharpey, and William Spottiswoode—were also on the Royal Society Council. Spottiswoode, general treasurer of BAAS for 1869–1870, also served as treasurer of the Royal Society from 1870 to 1878.

Therefore, many on the Royal Society Council would have been well aware of the situation with regard to Kew Observatory when, at a council meeting on 15 December 1870, a letter from BAAS dated five days earlier was read aloud, asking "what the desires of the Council of the Royal Society" were regarding the use of the Kew buildings. The council deferred the matter until 19 January 1871, when it appointed a special committee to discuss the BAAS proposal. In addition to the presidents and officers of both societies (thus including Sabine as president of the Royal Society), this committee included Galton, Gassiot, Alexander Strange, John Tyndall, Charles Wheatstone, and Alexander Williamson—every one of whom was also on the BAAS Council. The committee was given power to co-opt additional members: Warren De La Rue and William Grove were duly added on 16 February.[10]

Gassiot—who, along with Sabine, had been instrumental in starting instrument standardization in 1850, thus making it much harder to close down Kew—seems to have taken the initiative well before the committee met on 28 March. Balfour Stewart, George Airy, Humphrey Lloyd, William Henry Sykes, Thomas Romney Robinson, William Thomson, and the elderly John Herschel all wrote letters to Gassiot expressing their views on whether the Royal Society should take over the management of Kew Observatory. Their letters were clearly in response to solicitations from Gassiot. A letter from Gassiot to Herschel, dated 13 February, asks Herschel's "opinion as to the advisability of the Royal Society obtaining possession of the building with the view of ultimately continuing the Magnetic Observations Verifications of instruments, &c.." Gassiot wrote an almost identical letter to Airy the same day.[11] All the responses are dated mid-February 1871, with the exception of the letter from Balfour Stewart, which is dated 8 November 1870, just a month after he had taken up his new post at Manchester, suggesting that Gassiot had been sounding out opinions for at least three months.[12]

Most of the replies strongly supported keeping Kew Observatory going in some form. Balfour Stewart said he believed that "it would be a *very great misfortune*" if the Kew magnetograph work were terminated after 1872 because, true to his research interests in sunspots and their influences, he saw a need for "a more intimate comparison" between sunspot frequency and terrestrial magnetic disturbances, which rendered parallel magnetic and solar observations essential. Also, according to Stewart, differences in readings between observatories suggested that locality was important, so if Kew were to be given up, magnetic observations comparable with the Kew series could not be made elsewhere. Stewart also mentioned eleven observatories worldwide where instruments on the Kew design had been set up, "all of which would suffer were Kew discontinued." Humphrey Lloyd professed to have no strong views as to whether the Royal Society should start running Kew, though he generally felt that this type of work would be better done under the Royal Society than under BAAS. William Sykes—formerly of the East India Company, for which some of the earliest standardization work at Kew had been carried out in the 1850s—expressed disappointment that BAAS proposed to give up the observatory but strongly supported continuing it, especially standardization, under the Royal Society. Robinson, to whose design Robert Beckley had built the anemometer on the Kew dome, thought "that it would be a great loss to British science" were Kew to be given up and that if the standardization work were stopped, the need for another such laboratory "would be soon imperatively required by Physicists." William Thomson almost exactly echoed Robinson's sentiments in writing that closing Kew "would be a national calamity" and that the observatory had for the first time offered practitioners in the natural sciences a place for accurate observational work. Airy, predictably, recommended discontinuing the self-recording magnetic instruments at Kew on the grounds that these had been "introduced by me" at Greenwich in the 1840s (no mention of Francis Ronalds) and that duplicating this system at Kew was "an idle expense." But even Airy said that it was "very desirable that the Royal Society should have possession of the Kew Observatory." He approved of continuing Kew "for such purposes as were indicated in the original proposals for making use of Kew Observatory"— that is, improving magnetic instruments and the planning of "distant" (overseas) magnetic observations.[13]

The one dissenting voice among all these largely positive responses was that of John Herschel. Just as in 1850, when he had expressed the view that "it should most earnestly deprecate" the Royal Society for it to permanently maintain any observatory or institution, he now responded to Gassiot, "I should not feel very confident in recommending the Royal Society, as a body, to take on itself the duty of working *any* permanent scientific establishment." Herschel offered the same rationale for this view as he had in 1850: that supporting scientific institutions was not the Royal Society's mission, which in his view was rather to promote and manage science and to see that worthy scientific work was published and rewarded. On magnetism and meteorology, he took the same view as his old friend Airy: that both these sciences were now firmly established at Greenwich and so, by implication, it was not necessary to keep them going at Kew. As for solar photography, Herschel thought that this should be done by private individuals. In the version of his letter printed in the council minutes, Herschel was noncommittal as to whether instrument standardization was important enough to outweigh his general objection to the Royal Society taking over Kew.[14] Yet in a rough draft of the same letter preserved in Herschel's papers, he was dismissive of the idea that Kew should even be doing standardization, suggesting that this would best be done at the society's London headquarters or else should be taken up by the Board of Trade.[15] We do not know when, or why, Herschel changed his mind. Herschel was now nearly seventy-nine years old and physically frail, but while it is easy to suggest that he was "in decline" as an authoritative spokesman for science,[16] his mental capacity was still good. He kept an interest in astronomical developments—not least in sunspots, as is evidenced by his correspondence that same month with George Whipple at Kew in which he compares some Kew solar photographs with his own observations of the sun.[17] His response to Gassiot cannot be dismissed as that of an old man out of touch with the cutting edge of scientific research.

Herschel's reservations notwithstanding, the consensus among the seven leading figures in British physical sciences who responded to Gassiot was broadly in favor of the Royal Society taking over the running of Kew Observatory. The issue was made more complex when at the 16 March council meeting, a letter from Gassiot was read in which he made an offer of securities worth £250 per annum for the Royal Society to maintain Kew as "a Central Magnetical and Physical Observatory," this sum to be supplemented if the council deemed it insufficient to run the magnetic observations. Only a brief summary of Gassiot's letter was read at the meeting, but in the manuscript version preserved in the Royal Society's archives, Gassiot specified that this annual sum, nearly half of the £600 currently being voted on by BAAS for maintaining Kew, was not intended to support meteorology, instrument standardization, or experiments by private individuals—all of which, as he noted, were funded from other sources. This substantial offer would reduce—though not eliminate—the financial burden that running the observatory would present to the Royal Society: in Gassiot's own words, it would impose just "a very moderate charge" on the society's income.[18] Gassiot seems to have contemplated this offer for some time: the letters to him from Herschel and other senior figures, read later on during the 16 March meeting, in addition to his nearly identical solicitations to Herschel and Airy, suggest that he was canvassing their opinions before parting with his money—though neither of these latter communications give any hint that Gassiot was contemplating making any donation.

The special committee that met on 28 March 1871 was chaired by Spottiswoode, the Royal Society's treasurer. Among the thirteen Fellows present, six were also members of the BAAS Kew Committee; these included De La Rue, Galton, and Wheatstone, all stalwart supporters of the observatory under BAAS. Gassiot and Sabine did not attend, even though both had been invited onto the committee. It is possible that both wanted to distance themselves from this meeting—Gassiot because he wanted to avoid a conflict of interest now that his offer of a substantial donation was on the table, Sabine because he might have anticipated the result. The terse minutes of the meeting record how the committee rejected the idea of the Royal Society taking on the running of Kew Observatory, even with Gassiot's financial assistance. Two members of the committee put forward motions. First, Alexander Strange suggested that the committee was "not prepared to recommend the Council to undertake the responsibility of the maintenance of the Establishment." This motion was accepted by the committee. Then, the mathematician Thomas Archer Hirst put a motion rejecting Gassiot's offer. The only recorded dissent at this meeting was over the wording of this second motion, which was amended twice. The words "hoping that some other mode of giving the same generous assistance to the maintenance of the magnetical observations at Kew will suggest itself to him [Gassiot]" were omitted. In the final version, accepted at the meeting, the committee expressed its regret that it did not see in what way it could recommend that the council accept Gassiot's donation.[19]

Some of the committee's objections to the Royal Society taking on the observatory might have been for financial reasons: £250 was simply not enough to cover the current running costs of Kew, even if meteorology, standardization, and experiments could somehow be dropped from the program of work. Some other possible motives can be discerned by examining the backgrounds of the individuals who put forth the motions. At the 1868 BAAS meeting, Strange had presented his paper titled "On the Necessity for State Intervention to Secure the Progress of Physical Science," which had started the chain of events leading to the setting up of the government's Devonshire Commission to look into state provision for science education and research. As the original instigator of the lobby for greater state support of laboratories, he might well have felt that Kew was precisely the kind of institution that should be supported by central government and not private donations in the traditional manner. Also, Hirst's motion rejecting Gassiot's offer was seconded by John Tyndall, materialist and leading opponent of Balfour Stewart's Christian cosmology. Although Stewart was now at Manchester, he was still nominally superintendent of Kew, and Tyndall would not have wanted to see the Royal Society supporting what he would have seen as a hotbed of antimaterialist research.

The committee's resolutions were duly read out at the next Royal Society Council meeting on 27 April, with Sabine as chair, and it was resolved to consider the report at the next meeting. This took place on 25 May, again with Sabine present. This time Gassiot (figure 4.1) doubled his offer to £500 per annum, on the condition that the Royal Society maintained Kew "as a magnetical, meteorological, and physical observatory, with self-recording instruments" and that it be run by an unpaid committee of the Royal Society. Then a memorandum by Sabine was read, suggesting that if the unpaid committee stipulated by Gassiot were to be identical to the existing Meteorological Committee, whose members were also unpaid, this would overcome the technical difficulty (Herschel's objection, though Sabine did not name him) of the Royal Society supporting a permanent scientific institution. Under this scheme, there would be no worries about the Royal Society running such an institution because Kew would effectively be run by a government committee. Sabine did not need to remind the council that Gassiot's offer of £500 per annum covered most of the £600 currently being paid by BAAS and so removed most of the financial obligation from the Royal Society. If there were any objections at this council meeting, they are not recorded. The council resolved that the Royal Society's officers should, with the help of the society's solicitors, "prepare a scheme in reference to the Kew Observatory in accordance with Mr. Gassiot's views" and offer this to the council.[20]

What caused Gassiot to double his offer? Sabine must have prepared

FIGURE 4.1. John Peter Gassiot (1797–1877). Photograph by Wilson and Beadell. Image courtesy the Royal Society.

his memorandum in advance of the meeting, so he had presumably conferred with Gassiot and agreed on how to proceed. Moreover, three days earlier, Robert Scott of the Meteorological Office had sent Gassiot a detailed statement of the work being done by every member of the existing staff at Kew, as if Gassiot were asking Scott for a statement of what he was going to be paying for.[21] On 23 May, Gassiot seems to have ordered twenty-five copies of his proposal for forwarding to council members. In the same letter, he claims to have authorized De La Rue by letter prior to the 28 March meeting to increase his offer to £500, but that the letter never reached the committee.[22] This letter, if it ever existed, does not seem to have survived, nor does any evidence that Gassiot thought of offering £500 on 28 March. Gassiot's original letter of 13 March did allow provision for an increase, but it is unlikely that he had a 100 percent increase in mind. In any case, the original letter specifically excluded meteorology from his offer, whereas the new proposal was for the maintenance of "a magnetical, meteorological, and physical observatory." We are left wondering whether Sabine, ever the behind-the-scenes maneuverer, either twisted Gassiot's arm or made hints in this direction, especially as the committee's recommendations to reject Kew were not adopted at the next council meeting on 27 April but deferred until 25 May. Sabine might have been buying himself time for such negotiations.

However, between 27 April and 25 May there occurred a major event that might also have affected Gassiot's and Sabine's thinking. On 11 May John Herschel died. His burial at Westminster Abbey was practically an occasion of national mourning, as exemplified by William Thomson's presidential address to the 1871 BAAS meeting: "The name of Herschel is a household word throughout Great Britain and Ireland—yes, and through the whole civilized world."[23] More particularly for the fate of Kew Observatory, Herschel had been the one dissenting voice against the Royal Society taking over its management—yet also, in the eyes of elder statesmen of science like Gassiot, Sabine and De La Rue, the most venerable—whose views had carried such weight in the scientific world of the 1840s and early 1850s. With Herschel's death there was now no one left to object to Gassiot's proposals. Gassiot would have anticipated that the new proposals read out on 25 May would have been more attractive to council members because they would now relieve the Royal Society of most of the financial burden and would even technically relieve it of the practical responsibility of running the observatory. Yet under Gassiot's plan, Kew would still be run under the aegis of the Royal Society, something that Herschel had objected to even in principle. We have no way of proving that Herschel's death tipped the balance of Gassiot's mind, or the minds of others on the council, in favor of the revised proposal, but apart from Gassiot's unsupported claim, there is no evidence that Gassiot made any moves toward doubling his offer before 25 May.

Now that agreement had been reached, the council lost no time in implementing Gassiot's proposals. The day after the 25 May council meeting Spottiswoode, the treasurer, reported that he had instructed the Royal Society's solicitor, Charles Few, to talk to Gassiot's solicitor and draw up the heads of an agreement.[24] On 3 June, Gassiot gave Robert Scott instructions as to the general financial arrangements for the observatory: that its £400 annual allowance from the Meteorological Office should carry on as before, as should the salary of the superintendent (£200). With regard to any overtime to be paid to the assistants, Gassiot was ever the shrewd businessman: "We must take care not to commence with too high figures, as it is at all times difficult to reduce."[25]

The general terms of the agreement were presented to the council at its next meeting on 15 June. Gassiot was to present the Royal Society with £10,000 worth of securities, on trust, the income from which was to be used to run Kew Observatory and its work. The observatory and the trust income were to be managed by a committee appointed by the Royal Society Council. Yet although this committee's services were to be gratuitous, "like those of the present Meteorological Committee nominated at the request of Her Majesty's Government," neither the council minutes nor the trust deed itself specified that the new committee's membership was to be identical to that of the Meteorological Committee, as in the proposal outlined by Sabine on 25 May.[26] In the years after 1871, the membership of the new committee—still known as the Kew Committee—would cease to be identical to that of the Meteorological Committee. This suggests that Sabine had merely used a promise that the new Kew Committee would be identical to the Meteorological Committee in order to sugar the bitter pill of the Royal Society taking on the observatory's management—even after John Herschel, the most distinguished objector to this arrangement, was no more. It is very possible that many of the older Fellows of the Royal Society in 1871 would have shared Herschel's views on the society directly managing a scientific institution.

The final trust deed was sealed at the 29 June council meeting. At the same meeting, the council appointed the new Kew Committee to run the observatory under the Royal Society. The committee's membership—

De La Rue, Francis Galton, Gassiot, Admiral George Henry Richards, Sabine, Colonel William Smythe, Spottiswoode, and Wheatstone[27]—was indeed initially identical to that of the Meteorological Committee, so Sabine's promise was fulfilled to begin with, even though it carried no legal weight. Also striking, however, is that of these eight members of the new committee, six had been on the final Kew Committee appointed by the BAAS Council on 5 November 1870[28]—further suggesting that there had been broad agreement between BAAS and the Royal Society to transfer responsibility for the observatory. Management of the observatory formally passed to the Royal Society Kew Committee on 2 August 1871, when the BAAS Council, meeting at Edinburgh during that year's BAAS annual meeting, declared that the association could "give up possession of the Kew Observatory at once."[29] In his presidential address at the same BAAS meeting, William Thomson praised "the magnificent services which it [Kew Observatory] has rendered to science" and noted that the observatory now had "a permanent independence" thanks to "the noble liberality of a private benefactor, one who has laboured for its welfare with self-sacrificing devotion unintermittingly from within a few years of its creation." Yet, unsurprisingly in a presidential address, which one expects to be celebratory, Thomson skipped over any reference to BAAS being unable to continue supporting Kew and the complicated story as to how it came within the control of the Royal Society.[30]

Gassiot's donation was not quite the same as, for example, the Mond bequest to the Royal Institution, given by a businessman who had made his fortune in the chemical industry and who believed in the importance of laboratories to that industry.[31] Mond ultimately intended his donation to benefit his industry and increase its profits. Unlike Mond, Gassiot had become rich through activities far removed from those that he was now endowing. Nor was the Gassiot trust an instance of a businessman trying to buy his way into respectability by endowing a scientific institution, as was starting to become fashionable in the United States in the 1870s.[32] Gassiot had helped to set up the standardization program at Kew and had served as chairman of the Kew Committee since 1853, so he had a close and direct interest in the work of Kew Observatory, as he was proud to admit: "I need scarcely say that it has afforded me much pleasure, to have had it in my power, through the Royal Society, to assist in maintaining an Establishment with which I have, for so many years, been connected."[33] While Sabine was effectively the director of research at Kew, and might well have persuaded Gassiot to endow the observatory,

Gassiot clearly also had an ongoing personal interest in its work. The Gassiot trust therefore seems closer to being a case of a devotee of science privately funding research in which he had an interest, in the traditional Victorian manner. It surely also stands as a dramatic exception to Roy MacLeod's assertion that the Devonshire Commission's calls for greater state support were given greater urgency because "private philanthropy in support of scientific research was nowhere to be seen."[34]

AIRY AND KEW METEOROLOGY, 1871–1872

On 27 October 1870, barely a month after BAAS ratified the decision to relinquish responsibility for Kew Observatory, Sabine announced his intention of resigning as president of the Royal Society, effective 30 November 1871. By the latter date, he would have held the office for ten years. He was succeeded by his old adversary in the controversies between Kew and Greenwich, George Airy. There is no evidence that Airy had anything to do with Sabine's resignation. By the late 1860s, Sabine had been under increasing pressure, due to widespread dissatisfaction among Fellows toward his leadership.[35] A candid pamphlet by his friend Gassiot describing the events leading up to Sabine's resignation does not implicate Airy in any way.[36] According to Walter White, the Royal Society's full-time assistant secretary, who had intimate knowledge of Royal Society politics, Sabine's intention was that his successors would be from the aristocracy: first Lord Salisbury for two years, then Laurence Parsons, the 4th Earl of Rosse.[37] As Ruth Barton has persuasively argued, it was the influential "X-Club"—among whose nine members were Huxley, Spottiswoode, and Hirst—that was instrumental in nominating Airy for the presidency in 1871. The X-Club, which was keen to reduce the influence of the clergy and aristocracy in scientific societies, wanted the Royal Society to be led by an eminent, working scientist who, preferably, did not want to remain president for too long. Airy, as head of the world-famous Greenwich Observatory, fit this bill perfectly.[38] White records that in March 1871, Spottiswoode (an X-Club member) and George Gabriel Stokes (secretary of the Royal Society) visited Greenwich to offer Airy the nomination, that "he accepted without reserve," and that the nomination was unanimously supported by the council.[39]

Very soon after being elected president, Airy began an attempt to transfer to Greenwich the responsibility for the meteorological observations then being done at Kew as part of the Meteorological Office's system of observatories.[40] On 11 December 1871, less than two weeks after his election, Airy wrote to Meteorological Office director Robert Scott, claiming innocuously that "in my new position in connexion with the Royal Society, there has come before me general mention of the Kew Observatory and of its connexion with the Meteorological Office." He asked Scott how much the meteorological observations at Kew and also the reduction and printing of the results from the other self-recording stations were costing the government. In addition, he signified his wish to visit Kew.[41] Scott replied that the government was paying £250 per annum for the Kew meteorological observations. As soon as he had received Scott's letter, Airy arranged with Samuel Jeffery, superintendent of Kew Observatory since August 1871 (see the following), to visit Kew on 19 December. Airy's visit seems to have been a fact-finding mission, for he followed it up three days later by writing to Jeffery with some technical questions.[42] By 6 January 1872, Airy had written a paper to be circulated to council members prior to the next meeting. It pointed out that while the government was spending £250 a year on maintaining Kew as one of the self-recording meteorological stations, not far away the Meteorological Department at Greenwich, established several years before that at Kew, was "more complete in its equipment than the Kew Observatory, at least equal to it in the excellence of its instruments, and under the most careful daily superintendence." Airy thought it wrong, therefore, "still to load the Government with this unnecessary expense" and proposed that procedures should immediately be put in place to transfer to Greenwich the observations currently being done at Kew under the Meteorological Committee.[43]

Airy sent his paper to George Gabriel Stokes, Lucasian Professor of Mathematics at Cambridge. Stokes, described by David Wilson as "one of the great administrators of Victorian science,"[44] had been a secretary of the Royal Society since 1854 and so had been well aware of the positions of Kew Observatory and the Meteorological Office with respect to the Royal Society ever since the establishment of the Meteorological Department of the Board of Trade in 1854. Also, in 1851 Stokes had been awarded a sum of money from the Royal Society government grant to do some experiments of his own at Kew. His initial reply to Airy was friendly, though he urged caution, advising that Airy's proposal should be sent to the Kew Committee before the council decided on it. Then, four days later, Stokes wrote again to point out that the Kew observations came under the responsibility of the *Meteorological* Committee, whose authorization was "quite distinct" from the Kew one—an example of how the two committees, though identical in personnel for the time being, were legally different. The Meteorological Committee reported to the Board of Trade, not the Royal Society, and so Stokes now thought it "hardly proper" for the Royal Society to be questioning how a department of the Board of Trade was being run. Airy went ahead with his proposal anyway at the council meeting on 18 January 1872. Very near the start of the meeting, he raised the question of whether it was worth continuing the Kew meteorological observations at government expense, while "equally efficient" observations were "and have long been" done at Greenwich. The minutes merely record that after "some discussion," the matter was not pursued and that Airy announced his intention to follow it up directly with the Meteorological Committee.[45]

The Meteorological Committee, still chaired by Sabine, was by this stage well aware of Airy's renewed interest in Kew, for Scott had almost immediately informed the committee of his correspondence with Airy back in December.[46] Three days before the January Royal Society Council meeting, Scott bluntly reminded Airy of the limits of his jurisdiction by quoting a parliamentary paper, which pointed out that the Royal Society merely nominated the membership of the Meteorological Committee and had no connection with the Meteorological Office, thus fortifying the committee's position with legal sanction.[47] True to his announcement at the Royal Society, Airy wrote to the Meteorological Committee on 22 January, enclosing the same paper that he had circulated to the Royal Society Council. Airy was careful to note that he was sending this at the suggestion of the Royal Society's officers, implicitly denying any personal motive.[48] At its next meeting, the Meteorological Committee considered "various drafts" of a response to Airy. In the version sent, Scott again deployed the weapon of legal sanction in reminding Airy that the system of meteorological observatories "was adopted by Her Majesty's Government" and was a matter for the Board of Trade. Moving the Kew meteorological observations to Greenwich would mean placing them under a different government department—the Admiralty—over which the Meteorological Committee had no control. In addition, claimed Scott, Kew had to be one of the self-recording observatories if it were to work properly as a nerve center for the other stations scattered across the British Isles. At the end of his letter, Scott called Airy's bluff, informing him that the committee "will be ready to advise the Board of Trade . . . if they should be consulted in the matter."[49]

Airy does not seem to have approached the Board of Trade on his own, likely because he presumed that the board would throw the question straight back to the Meteorological Committee—or, even worse, its chairman, Sabine. Yet he did propose that the Royal Society Council follow this strategy collectively. On 11 April he announced that he would put three motions before the council at its next meeting: first, that it was within the "competency" of the Royal Society to enquire into whether it was necessary for Kew to remain the central observatory and make representations to the Board of Trade if need be; second, that responsibility for the meteorological observations now being made at Kew should be transferred to Greenwich; and finally, that a copy of the second motion should be sent to the Board of Trade. Airy put the first of these motions near the end of the 18 April council meeting, but no one seconded it and so Airy did not see it as worthwhile to move the second and third resolutions.[50]

No further moves by Airy to transfer the Kew meteorological observations to Greenwich are recorded in any minutes or correspondence for the rest of Airy's tenure as president of the Royal Society. Airy had no jurisdiction over the Board of Trade, a different department from Greenwich's governing body, the Admiralty. Seeing that there was no way of ever persuading the Meteorological Committee, at least for as long as Sabine remained its chairman, Airy used his position as president to seek the Royal Society Council's authority to persuade the Board of Trade. Yet neither could the Royal Society simply tell the Board of Trade what to do. As Walker has noted, we may never know the unrecorded machinations behind the scenes at the council,[51] but it is reasonable to imagine Airy presuming that, as a majority of its members (eighteen are recorded as present on 18 April) were not involved with the Meteorological or Kew Committees, he had a chance of persuading enough of them to take his side. Yet it is also plausible that the council members would have agreed with Stokes that it would be "hardly proper" for the Royal Society to be questioning the running of a department of the Board of Trade.

As several historians have shown, Airy had long expressed the view that the burden imposed by Greenwich, and science in general, on the public purse should be minimized as far as possible. He had repeatedly aired the opinion that only observations with a utilitarian purpose should be done at Greenwich, while those of an experimental nature or with no practical applications should be carried out by private individuals at their own expense, or perhaps with occasional grants from funds such as the Royal Society government grant.[52] It is easy to envisage, therefore, Airy taking the same attitude with regard to the Kew meteorological observations, which he saw as an unnecessary duplication of work at public expense. But if he simply wanted to avoid duplication, he might have indicated that he wished to discontinue the Greenwich meteorological work, now that a perfectly good system was running at Kew. Similarly, Airy's attempts in the 1860s to stop the Kew *magnetic* observations could not have been driven by a desire to save public money, as these were privately funded by BAAS. A wish to keep hold of the staff who ran the Greenwich magnetic and meteorological department—especially James Glaisher, one of his most loyal members of staff as well as the most publicly known—would surely also have been a factor in Airy's desire to continue the meteorological observations at Greenwich. Yet one suspects that a major motive of Airy's in his attempts to wrest control of the magnetic and meteorological observations from Kew was to keep his place at the top of the hierarchy of observational astronomy and its allied sciences. To Airy, rival establishments like Kew were usurpers. Airy's attempt to take over the Kew meteorological observations can be seen as a failed attempt to put down this longtime usurper.

THE KEW PHOTOHELIOGRAPH: AIRY'S SUCCESSFUL COUP, 1872–1873

Airy was much more successful in transferring to Greenwich another branch of research for which Kew had become famous by the early 1870s: solar photography. The Kew and Meteorological Committees did not create the same difficulties over this because the Kew Committee was already planning to terminate the program once a full ten years of continuous solar photographs at Kew had been completed in 1872. As with the meteorological observations, Airy's motive was partly to redress the balance of power between Greenwich and Kew. However, Airy also seems to have been motivated by a desire to forestall moves by the ongoing Devonshire Commission, and in the Royal Astronomical Society, to set up a new solar physics observatory under the direction of Norman Lockyer. Related to this, the story of Airy's successful acquisition of the Kew photoheliograph suggests that he had ulterior motives in accepting the presidency of the Royal Society.

During his visit to Kew in December 1871, Airy enquired about solar photography in addition to the meteorological work there.[53] Then, on New Year's Day 1872, Airy wrote to De La Rue in a private capacity, saying that he was "sorry to hear" that De La Rue's solar photography project was coming to an end so soon, something that Airy must have been aware of for some time. He asked whether part of the Gassiot money could be diverted from magnetism to continuing the solar work, knowing that this would never happen for as long as Sabine remained in control of Kew. Then, in a postscript, he remarked that "I set great value on the continuation of the sun-pictures, and regret that I cannot take them up here." Airy seems to be hinting here that he wanted them for Greenwich. De La Rue's enthusiastic response the next day strengthens this possibility. He expressed the wish that the solar observations should be carried out on a permanent basis and should be funded by the government: "I wish very much that solar photographic observations could be made the business of a Government Establishment"—of which Greenwich was the only example for astronomy in England. The cost, said De La Rue, would only be around £200 per annum. He concluded with the comment that "if ever meteorology is to be placed on a scientific basis that [sic] it will have to be studied in connection with solar phenomena."[54]

Airy and De La Rue were old friends by the early 1870s, as is clear from their reciprocal New Year's greetings in the preceding exchange of letters. To Airy, De La Rue must have been a shining example of how he believed astronomy outside Greenwich should work, with new fields such as solar photography being pioneered by wealthy devotees of science such as De La Rue, who funded their research from their own resources. De La Rue's comment about meteorology must also have been music to Airy's ears, not least because Airy himself felt that meteorology needed a stronger theoretical footing before it could properly be called a science. De La Rue was a member of the Board of Visitors, Greenwich Observatory's governing body, and it may not be surprising that, at a meeting of the board on 1 June 1872, he proposed that "the time has arrived when it would be for the advantage of Science that continuous photographic and spectroscopic records of the Sun should be made at the Royal Observatory."[55] Airy soon followed up this proposal. In October, Airy asked the Royal Society Council's permission to borrow the Kew photoheliograph, on the grounds that the government had just spent a great deal of money on several new photoheliographs for the 1874 transit of Venus and so should not be asked to fund an additional instrument. Once again, Airy used the pretext of saving public money to advance his interests—in this case, transferring the photoheliograph to Greenwich so that no rival could use it. The council willingly gave its assent to Airy's request: solar photography was now redundant at Kew and had nothing to do with the Meteorological Office.[56] With permission granted, Airy lost no time in moving the photoheliograph to Greenwich. Just a week after the 31 October council meeting, an agent of Airy's had called at Kew to examine the instrument. By June 1873 the photoheliograph had been placed in a dome at Greenwich, and it was in regular use from April 1874.[57]

That Airy should so artfully negotiate the transfer of the Kew photoheliograph to Greenwich might appear inconsistent with his highly utilitarian stance on Greenwich's role, as sunspot observation was surely far removed from timekeeping and navigational astronomy. Indeed, Rebekah Higgitt has argued that Airy had to be pressed by the Board of Visitors into diversifying into the new type of astronomy.[58] Simon Schaffer and Roger Hutchins have each suggested that Airy accepted the introduction of solar photography because photography was a form of automation that could control observer error, similar to his device for timing star transits.[59] But why institute solar observation at all, automated or not? Because of its alleged connections with terrestrial magnetism and meteorology, Airy might have considered sunspot research as being no less utilitarian than the magnetic and meteorological observations that he had been running at Greenwich for thirty years. He might have inferred this from the assertion in De La Rue's letter of 2 January 1872 that if meteorology were ever to be considered scientific, connections had to be made between it and solar activity. Acquiring the photoheliograph made perfect sense as an extension to the existing magnetic and meteorological department. For as long as the instrument was operated by his friend De La Rue, Airy was prepared to tolerate it being run at Kew. But now that it was redundant, he did not want it going into anyone else's hands and so he had to have it. As the "National Observer," Airy saw solar astronomy as his prerogative as much as magnetism and meteorology.

By the time the photoheliograph was in regular use at Greenwich, Airy was no longer president of the Royal Society. He stepped down from the society's anniversary meeting effective November 1873, after just two years in the post; indeed, he had announced his intention to resign at the previous year's anniversary meeting and confirmed his decision at a council meeting in December 1872.[60] His presidency was the shortest since William Wollaston had held the post for just five months in 1820 after the death of Joseph Banks. Yet it is striking that Airy made his first recorded moves with regard to the Kew meteorological observations on

11 December 1871, less than two weeks after he was elected president, and he made his first enquiries about the Kew photoheliograph just over a fortnight later. Historians have generally accepted the official reasons Airy gave for his resignation in his autobiography and in his final presidential address: that the position involved too much work; that he was based too far from central London, where the president of the Royal Society frequently had to attend meetings; and also "a difficulty of hearing, which unfits me for effective action as Chairman of Council."[61] Yet it seems scarcely plausible that someone as astute as Airy, a Fellow of the Royal Society since 1836 and with long experience of the society's affairs, was not aware of the nature of the president's role before he took it on. In fact, in December 1870, just six weeks after Sabine announced his resignation, Airy began a letter to De La Rue: "Since our last conversation, I have thought repeatedly on the Presidency of the Royal Society. And my feeling is, that it is encumbered with many difficulties." The "difficulties" recognized by Airy included that of managing his work as Astronomer Royal so that he could devote sufficient time and attention to the presidential duties, as well as "the absorption of time and strength by attendance at Councils and Meetings," the fact that he lived "an hour's journey" from the Royal Society's premises, and also his "slowly increasing deafness."[62] Thus Airy indeed had few illusions as to the amount and nature of the work involved if he accepted the office of president: in 1870, he had anticipated all the reasons that he eventually gave for resigning.

In addition, that Airy wrote to De La Rue as early as December 1870 shows that he was seriously considering taking on the post of president very soon after Sabine's resignation. Moreover, his opening words, "since our last conversation," show that verbal discussions about the possibility had already been taking place between Airy and De La Rue. All this suggests that Airy had an ulterior motive in becoming president. Now that Kew Observatory was to be run by the Royal Society, Airy might have seen the presidency as an opportunity to at last transfer back to Greenwich some of the balance of power that Sabine, in Airy's eyes, had stolen for Kew. Further evidence for Airy thinking along these lines is contained in a letter by Balfour Stewart dated soon after the announcement of Airy's resignation: "I hear that Airy has twice tried to stop the Meteorological Committee and no doubt Kew Observatory also but I fancy it was seen by the Council that his motives were not pure so that he was snubbed and has expressed his intention of resigning."[63] For as long as it offered the possibility of lowering Kew Observatory to what he saw as its correct place in the hierarchy, Airy might have regarded the increased workload that came with the presidency worthwhile. But now that the photoheliograph was securely in his possession—even though his coup attempt over the Kew meteorological observations had been "snubbed" by the council—Airy clearly saw no point in carrying on as president. His nomination for the presidency in 1871 may have suited the X-Club, but for Airy to have allowed his name to go forward it must also have suited Airy.

However, Airy is likely to have had a second ulterior motive in acquiring the Kew photoheliograph that would have a much greater significance for understanding the ongoing debates in the 1870s about state support for science. The issue of government-funded observatories and laboratories was not addressed until the Devonshire Commission's eighth and final report, published in 1875, but the relevant hearings took place in the spring and summer of 1872, when Airy and De La Rue both testified. The Devonshire Commission also interviewed Alexander Strange, who had instigated the chain of events that had led to the commission being set up. The hearings were conducted by small panels of well-known scientific personalities, notably Thomas Huxley and George Gabriel Stokes, under the chairmanship of the Duke of Devonshire. Airy, De La Rue, and Strange were all questioned as to the possibility of establishing a new, state-funded observatory dedicated to the new astronomical physics, especially photographic and spectroscopic studies of the sun. Strange expressed the belief that such an observatory, if established, had to be independent of Greenwich.[64] De La Rue took a more ambivalent position: a new observatory, he said, should be under the aegis of Greenwich and come under the Astronomer Royal, but it need not necessarily be built at Greenwich.[65] Airy, on the other hand, did not believe that observatories for open-ended, nonutilitarian research should be established at cost to the public purse—though he did express the view that regular solar observations could be done at Greenwich. Airy specifically said that daily solar photographs like those being done until recently at Kew should be publicly supported, although he did not refer to his correspondence with De La Rue earlier in 1872, which had set in motion the transfer of the Kew photoheliograph to Greenwich.[66]

By the time Airy had presented his views to the commission, the idea of setting up a state-funded observatory for astronomical physics had become a contentious issue in the Royal Astronomical Society. At a meeting of the society on 12 April 1872, Strange presented a paper provocatively titled "On the Insufficiency of Existing National Observatories." Strange strongly advocated the establishment of a new observatory, separate from Greenwich, dedicated to solar research and stellar spectroscopy, complete with a laboratory for chemical analysis of spectroscopic results, plus a series of other observatories across Britain's imperial possessions. In this regard, he expressed particular anxiety at the recent termination of the Kew photographic observations: the sun was no longer being systematically monitored anywhere in the British Empire, and it would be "an evil" if this situation were to continue. As Barbara Becker has acknowledged, Strange's motive was likely to rally support for his testimony to the Devonshire Commission, which held its hearings about observatories very soon after this RAS meeting.[67] Strange was using this paper in order to guide the commission toward his own aims for greater state involvement in research such as solar physics.

Strange's paper was followed by a discussion in which Airy expressed reservations about Strange's proposals. He asserted that government observatories should, first and foremost, concentrate on long-term, utilitarian work such as the meridian observations for navigational purposes being done at Greenwich "because Governments want to have something to show for their money." Making particular reference to the solar work recently terminated at Kew, Airy remarked that a government observatory "could not go groping about for the causes of solar changes, such as planetary conjunctions and the like, but must occupy itself in making definite series of observations." In Airy's view, "it was the place of a Government not to establish philosophical institutions, but working bodies." However, Airy was flexible as to what new work Greenwich might take on in the future: "If theory or antecedent observation show a worthy object, the necessary investigation should be sanctioned."[68]

In the weeks after the April 1872 meeting, the proposed observatory for astronomical physics became the subject of heated debates at RAS Council meetings, with Strange and Lockyer passionately in favor of the new observatory and most of the other council members firmly against it. The controversy culminated at the council meeting on 29 June 1872, which voted in favor of sending to the Devonshire Commission a memorandum recommending that no separate observatory for solar physics be established, "especially as they have been informed that the Board of Visitors of the Royal Observatory at Greenwich, at their recent meeting, recommended the taking of Photographic and Spectroscopic records of the Sun at that Observatory."[69] This was the first public announcement that solar photographic observations were to be established at Greenwich, as agreed by the Board of Visitors earlier that month and as privately arranged between Airy and De La Rue at the beginning of 1872. This move seems to have been kept secret until 29 June, for no mention of it was made in any publication before then. Only on 12 July, when it was his turn to testify, did De La Rue mention the board's decision to the Devonshire Commission and remark that "I believe it is in contemplation to establish such a series of observations" at Greenwich[70]—knowing full well that these had been "in contemplation" since January.

The narrative of Airy's private moves to acquire the Kew photoheliograph and then the controversies over the proposed solar physics observatory reads very much like the sequence of events in 1840, when Sabine's faction had applied to the government for a separate magnetic and meteorological observatory and then Airy, when he had come to hear about it, had punctured Sabine's plan with a proposal for his own extended magnetic and meteorological department at Greenwich, at a substantially lower cost to the public purse. Now, in the 1870s, Airy once again put a stop to a plan for a rival observatory by not only offering to take on solar work at Greenwich but this time arranging the transfer of the photoheliograph with De La Rue months before the plan was announced.[71] It is likely, then, that these moves by Airy were not only one of the last acts in his long rivalry with Kew but were also in anticipation of the Devonshire Commission. By early 1872 Airy, as president of the Royal Society and having an intricate web of connections in the London scientific world, would have known about the moves afoot to review public scientific institutions and would have been perturbed by the idea of Strange and Lockyer forestalling him with a separate observatory. Lockyer eventually received authority to establish his own solar physics observatory at South Kensington, but this did not happen until 1881, the year of Airy's retirement as Astronomer Royal, by which time daily solar photography was firmly established as part of the routine at Greenwich.[72]

THE WORKING OF KEW OBSERVATORY, 1871–1885

When compared to the high drama of the 1860s and early 1870s—the solar discoveries, the Sabine-Stewart confrontations, the handover to the Royal Society, and Airy's machinations—it is easy to think of the history of Kew Observatory between the early 1870s and mid-1880s as a long period of stability in which the observatory continued its existing meteorological and standardization work, in addition to remaining an important center for geomagnetic observations. It is certainly possible to gain this impression from reading Robert Henry Scott's 1885 history of Kew Observatory. Yet the period between the Royal Society taking over the observatory in 1871 and the publication of Scott's paper in 1885 was less stable than it seemed. During the mid-1870s, Kew went through a lean time in terms of funding and scientific output. Some key members of the Kew Committee—notably Sabine—withdrew from the scene, leading to a lack of leadership at the top. The observatory's fortunes improved from the later 1870s, but with this came a change of emphasis. Geomagnetism, although still important, became a routine and somewhat secondary aspect of the observatory's work, while meteorology assumed greater relative importance. But above all, instrument standardization became the dominant activity at Kew in these years. By 1885, standardization was Kew Observatory's largest single source of income. This section argues that financial pressures, as well as the changing makeup of its governing committee, forced Kew to become essentially a laboratory that served practitioners of science based elsewhere, rather than a place of research as the BAAS Council had at least partly envisaged it prior to 1871.

When Balfour Stewart resigned as superintendent of Kew Observatory in October 1869, Sabine and Gassiot offered his job to the former magnetic assistant Charles Chambers. In any event, however, Chambers stayed in his post at the colonial observatory in Bombay. Stewart remained nominally superintendent, even after taking up his professorship at Manchester in the autumn of 1870. The correspondence shows him running the observatory remotely, down to paying the assistants' salaries.[73] He continued in the post even after late November 1870, when he was severely injured in a train crash that left him house-bound in Harrow, Middlesex, until the spring of 1871. During these months his wife Katherine, known informally as "Katie," wrote to the staff at Kew on his behalf. For example, in January 1871 she wrote to the chief magnetic observer, George Whipple: "Mr. Stewart would like to hear how the magnetic work is getting on? . . . He is going on very well, he can pull himself from one side of the bed to the other."[74] By 1871, it might have looked as though Kew Observatory was being run by an extended family, for Katie Stewart was not the only woman helping to run the place. In June 1870, Elizabeth Beckley had married George Whipple. She had played an essential role in taking the sunspot pictures and was now measuring the surface areas of sunspots on the photographs.[75] Katie refers to her as "Lizzie" in her 27 January letter: from the familiar name she was clearly part of the Kew "family."

Lizzie's husband had, in fact, assumed the day-to-day running of the observatory after Stewart had left for Manchester. Scott's statement to Gassiot of the work being performed by each member of staff has "general supervision of the daily work of the Observatory" as first on the list of Whipple's duties. Whipple was also in charge of correspondence and finance, in addition to supervising the magnetic and meteorological observations.[76] After joining Kew in 1858 as a boy of fifteen, he had initially carried out meteorological observations before progressing to magnetic work. Since Chambers's departure in 1863, Whipple had been the most senior assistant at Kew: Scott's 1871 letter to Gassiot refers to him as "1st. Assistant." In 1871, too, Whipple was awarded the degree of bachelor of science after completing a University of London degree course.[77] Thus in addition to his lengthy experience of magnetic and meteorological observation, Whipple by 1871 had a university training. With hindsight, he might seem to have been another John Welsh or Balfour Stewart in the making and therefore a logical successor to Stewart when the latter finally resigned as superintendent on 27 June 1871.[78]

Yet on 3 July 1871, at its first meeting, the newly constituted Kew Committee of the Royal Society appointed Samuel Jeffery as superintendent. Jeffery was a complete newcomer to Kew, as he revealed in a letter to Whipple later that month in which he expressed his intention to visit Kew before starting his appointment "to become somewhat familiar with the daily routine."[79] Of all the superintendents at Kew between 1842 and 1910, Jeffery is very much the least well known today. As Savours and McConnell have pointed out, no obituary for him has ever been found. The Kew Committee gave no official reason for choosing him in preference to Whipple or anyone else. Like Chambers, Jeffery had served a long apprenticeship at a colonial observatory. He had begun work at the Rossbank magnetic observatory in Hobart, Tasmania—one of Sabine's original Magnetic Crusade stations—in 1840, initially as a volunteer, before serving as a paid assistant at Rossbank between 1842 and 1853. He was then director of Rossbank until the observatory closed in 1854. Following this, he seems to have suffered some years of unemployment before he joined the Meteorological Office in London in January 1869 as a senior clerk, assisting with reducing the data from the self-recording observatories.[80]

For all the relative obscurity of his career, however, Jeffery had one attribute that was essential for a post at Kew Observatory: he had never found himself on the wrong side of Edward Sabine. It is likely that Sabine was instrumental in recruiting him for Kew: in 1867, Stewart had remarked to Sabine that if a vacancy were to arise at Kew, "I will bear in mind what you say of Mr Jeffery."[81] Sabine might have been motivated by a wish to help out a fellow magnetic crusader now down on his luck. Yet it is also probable that Sabine—and others on the Kew Committee —chose Jeffery precisely *because* Whipple had the makings of another Balfour Stewart. After Stewart's disagreements with Sabine and Gassiot, the committee did not want another independent-minded researcher. It would have suited Sabine and his colleagues to have someone who, as a colonial observer and later a humble clerk at the Meteorological Office, was used to being in a subordinate position. Both Stewart and Whipple seem to have been surprised at the choice of Jeffery. Whipple might naturally have thought that the post would go to him. Also, now married and with a baby shortly due, he would have appreciated the increase in salary.[82] Whipple seems to have at least considered resigning, while Gassiot did "not like the tone of Stewarts [sic] letter, I suppose he is offended at yr. not accepting Mr W. as his successor."[83]

In its last months under BAAS, Kew Observatory had a complement of nine assistant staff, plus two temporary "supernumerary" assistants to help with the reductions of observations and two part-time assistants working with the photoheliograph.[84] Thanks to the Gassiot Trust, the Kew Committee did not have to lay off any staff—with the notable exception of mechanical assistant Robert Beckley, who had built the cup anemometer on top of the dome. In December 1871, he was made redundant on the grounds that the committee did not believe that there would now be enough work for a full-time mechanical assistant. The committee continued to pay him a retainer of £10 a year for any ad hoc design work that might be needed.[85] Beckley's dismissal was perhaps an early instance of how Kew after 1871 was becoming less orientated toward research: the Kew Committee now perceived that there would be less need to develop new instruments. In the same year, two further magnetic assistants started work at Kew to help with Sabine's magnetic reductions, though they were paid by the War Office since there was no longer any room for a magnetic office at Woolwich.[86]

Jeffery started work as superintendent of Kew on 2 August 1871, the official day of the observatory's handover to the Royal Society. The Kew

Committee's wish for a compliant director is evidenced by its simple and rigid definition of Jeffery's duties: first, to act as general superintendent of the observatory; second, to be director of Kew as the "Central Observatory" of the Meteorological Office; and third, to supervise the meteorological results.[87] Unlike Stewart, he did not become secretary to the Kew and Meteorological Committees; Scott filled both these roles instead. Yet from quite early in his time as superintendent, members of the Kew Committee began expressing dissatisfaction with Jeffery's performance. By the end of 1872, Scott was writing to Jeffery in exasperation: "I cannot understand how you mean that the max & min readings were 'beyond your criticism'[.] Kew exercises a supervision over every line & figure sent up from the observatories."[88] Jeffery was not expected to do research and experiments in the manner of his predecessor, but his position still involved running a multifunction observatory that did geomagnetic observations and large-scale testing of instruments as well as meteorology. Yet Jeffery seems to have simply continued what he had been doing as a clerk at the Meteorological Office, supervising the returns from the outlying observatories—and not always competently, as Scott's letter suggests. By the spring of 1874, the reductions of observations were seriously in arrears. This situation remained unresolved by the autumn of 1875, by which time other tabulations were months behind schedule as well. Jeffery also seems to have lacked the aptitude for managing people. In November 1875, when Jeffery apologized to Scott about the slow progress being made by one of his assistants in clearing the backlog of work, Scott had to remind Jeffery that "you are the best judge being on the spot. If . . . is not up to his work he should be dismissed *at once*."[89] Although Jeffery had briefly been superintendent at Rossbank, he had only ever had one assistant there and frequently worked on his own. At Kew he had to manage at least eight full-time staff.

That this situation was allowed to persist for so long seems to have been partly because the Kew Committee was itself in difficulties by the mid-1870s. After 1871, the committee remained dominated by the elder statesmen who had run the observatory under BAAS since midcentury: men like De La Rue, Gassiot, and Wheatstone, as well as Sabine. In 1875, Wheatstone died and Gassiot resigned due to ill health. Neither of these two was immediately replaced, causing the committee to shrink in size. After Spottiswoode resigned in 1873, Richard Strachey (a lieutenant-general with particular interest in Indian weather and climate) and Lord Rosse (son of the 3rd Earl of Rosse, who had served as president of the Royal Society from 1848 to 1852) were invited onto the committee,[90] but its meetings in the early 1870s were often attended by just three or four people, two of whom were Scott and Jeffery. Although Sabine nominally remained chairman until his death in 1883, he is last recorded as having attended a meeting of the Kew Committee on 18 December 1874. There are clear signs that the octogenarian was winding up his affairs by this time, including donating his library to Kew Observatory. By mid-January 1876, Sabine was "well in health but his mind is inactive now."[91] Of the "old guard" on the Kew Committee, De La Rue alone remained. He usually chaired committee meetings on Sabine's behalf, but even he had by now retired from active astronomical observation.

Therefore, a lack of leadership on the Kew Committee might well be a reason why Jeffery was allowed to remain superintendent while the backlog of work accumulated. Jeffery resigned at the Kew Committee meeting of 19 November 1875, after the committee had informed him that in the future the observatory would need to be headed "by a scientific man" with "special scientific knowledge."[92] It is indicative of the committee's hidden agenda in 1871 that it could then have appointed just such "a scientific man," George Whipple, to the post of superintendent. It is interesting that in the same letter in which he referred to Sabine's "inactive" mind, Scott was able to inform the head of another observatory that Jeffery was to leave Kew on 1 March 1876 and would probably be succeeded by Whipple.[93] It is possible that Scott and others on the Kew Committee had for some time wanted to dismiss Jeffery and replace him with Whipple but felt unable to do so without offending Sabine, because Jeffery was Sabine's protégé. But now that Sabine was safely out of the picture, they would have had less inhibition in persuading Jeffery to resign. The fact that the committee offered Jeffery £100 in gratitude for his efforts, gave him the option to leave before March 1876, and even agreed to buy his furniture for £28 greatly strengthens the idea that its members wanted Jeffery out of Kew as soon as possible. On Jeffery's departure, George Whipple immediately became acting superintendent; he was appointed full superintendent on 14 November 1876.[94]

The events of the mid-1870s mark an important watershed: the end of Sabine's leadership of the observatory. For the first time since 1842, the scientific agenda at Kew would no longer be dictated by Sabine. The decline and departure of Sabine might explain why geomagnetic work at Kew from the 1870s onward was less innovative than it had been in the 1850s and 1860s. The self-recording magnetometers remained practically operational throughout the period from 1871 to 1885, but less *research* was done with them. Jeffery had no recorded involvement with the magnetometer work; this was left in the charge of Whipple, as it had been before 1871. With Sabine gone and the Kew Committee's vice-chairman, De La Rue, increasingly elderly, overall supervision of the work at Kew fell to Scott, who as head of the Meteorological Office already had more than enough on his schedule. Geomagnetism was certainly not part of Scott's Meteorological Office duties, as he politely but firmly reminded the son of the Belgian astronomer and statistics pioneer Adolphe Quetelet: "Please remember if you write to me on Magnetical business to keep the letter separate from any communications on Meteorological business. This *office* takes no cognizance of Magnetism."[95]

One of the first recruits to the Kew Committee in the post-Sabine era was William Grylls Adams, James Clerk Maxwell's successor to the chair in physics at King's College, London. In 1879, Adams persuaded the Kew Committee to stop the long series of tabulations of magnetic curves in favor of comparing the curves themselves with those of foreign observatories, "with a view to the development of the theory of magnetic disturbance." This was another effect of the departure of Sabine and his unquenchable thirst for more and more data that had so exasperated Balfour Stewart in the 1860s. It also reflected Adams's own interest in precision measurement, not least in magnetism, which featured prominently in the physics syllabus at King's.[96] The international status of Kew as a center for magnetic observations and instruments remained undiminished into the 1880s. Magnetographs and dipping needles tested at Kew were regularly supplied to foreign observatories, such as Imperial Germany's prestigious new Potsdam Astrophysical Observatory in 1876.[97] Kew also sold standard forms to observatories for recording magnetic observations; among the customers for these in 1882 was the Cavendish Laboratory in Cambridge, which had been established in 1874. One of the Cavendish's earliest acquisitions was a magnetometer originally used at Kew. James Clerk Maxwell, the laboratory's first director, used this to train students in making precision measurements.[98] Yet all this work was in the service of other institutions or individuals and not for research of the type that had been done by Sabine. Far from being the command post of Sabine's Magnetic Crusade, Kew was now a kind of service regiment providing instruments and expertise for other magnetic projects.

After 1871, Kew remained the Meteorological Office's central observatory for land meteorology, coordinating, checking, and reducing the observations sent in by the six outlying stations: Falmouth, Stonyhurst, Glasgow, Aberdeen, Armagh, and Valencia (Ireland), each one equipped with self-recording instruments tested at Kew. The reductions were carried out by two junior assistants at Kew under the supervision of Jeffery and then, from early 1876, under Whipple. This was the "bread and butter" of Kew meteorology until the autumn of 1876, when the Meteorological Committee decided to move this work to the office's headquarters on Victoria Street, London, on the grounds that it would be more efficient to do the reductions in the same place where the office's quarterly weather reports were produced. As a result, one of the assistants doing the reductions moved from Kew to London. This slightly reduced the Kew salary bill, but the Meteorological Office reduced its annual allowance to Kew from £650 to £400, leading to a significant loss of income. In addition, one of the Kew assistants still had to regularly travel to the outlying observatories to check the instruments, and another assistant had to remain on standby to cover for absence at any of these stations.[99] More importantly, the change led to a major reduction of status for Kew. It was now just one of seven self-recording observatories reporting to the office—a far cry from Galton and Sabine's 1866 plan for the office to act as "a *branch office in London*" for Kew. The Meteorological Office was now taking the lead and Kew Observatory had to follow. The decision to transfer the work to London might have been a consequence of the disorganized state of Kew under Jeffery, but once again it took place neatly after the departure of Sabine. Now that Sabine was no longer actively on the scene, there was no compulsion to follow his cherished plan for Kew as the Meteorological Office's central observatory.

A further change in the balance of power between Kew and the Meteorological Office came in July 1877, when the members of the Meteorological Committee resigned en bloc and were replaced at the same meeting by a new "Meteorological Council." This was in response to a new Treasury inquiry into the Meteorological Office, instigated in November 1875, asking it to justify its £10,000 annual budget. As Anderson and Walker have both pointed out, this inquiry was somewhat internal: some of its members were on the existing Meteorological Committee and two of them (Galton and Thomas Farrer) had served on the original 1866 committee that had set up the current arrangements.[100] Perhaps unsurprisingly, the inquiry's report recommended that the Meteorological Office be kept going much as before, except that members of the new Meteorological Council should be paid, the total salary bill not to exceed £1,000 a year.[101] From 1877 also, Scott was no longer secretary to the Kew Committee and so was less actively involved in the daily running of the observatory. The secretary's role passed to Whipple, though Scott remained on the Kew Committee as an ordinary member. The main result of the report for Kew was that the Meteorological Council's membership was no longer the same as that of the Kew Committee—in contrast to Sabine's 1871 assurance that the two committees would be identical. In 1879, for example, the Meteorological Council was chaired by Henry Smith (Savilian Professor of Geometry at Oxford since 1860); George Gabriel Stokes was also a member. Neither of these two were on the Kew Committee at this time. That same year, some members of the Kew Committee, such as Richard Strachey and George Carey Foster (professor of experimental physics at University College, London, since 1865), were not on the council.[102] This is further evidence that making the two committees identical had been a ruse by Sabine to reassure Fellows of the Royal Society that they would not have to live John Herschel's nightmare of an observatory being run by the society. More importantly for the future development of Kew Observatory, members of the Kew Committee now had to take their instructions, where meteorology was concerned, from a more explicitly separate body. The separation of the two committees helped to lift the mask off the fact that Kew was no longer the Meteorological Office's nerve center but was an independent institution, relying on the Gassiot money plus income from standardization to keep going.

This shift in the balance of power between Kew and the Meteorological Office in the post-Sabine era did not mean that Kew meteorology lost any status in the outside world. The 1870s saw summaries of Kew meteorological observations being printed in newspapers such as the *Times* (at the editor's request) and the *Illustrated London News*.[103] After he became superintendent at Kew, Whipple began some innovative experimental work. In 1879, he began a two-year comparison of two types of screen for shading thermometers, one of them the famous "Stevenson screen" designed by British engineer Thomas Stevenson (father of the author Robert Louis Stevenson), the other by Heinrich Wild of the St. Petersburg Observatory in Russia.[104] Whipple took charge of a more ambitious experiment five years later, when the Meteorological Council granted the Kew Committee £40 to set up a system of two cameras 800 yards apart, connected by telegraphic cable, for photographing clouds simultaneously in order to measure their heights and their speeds and directions of motion. Simultaneity was ensured by connecting the two observers with the newly invented telephone. The photographic work began in July 1885, and over the ensuing weeks Whipple and his assistants made sixty-two measurements of the speeds and directions of cloud motions.[105]

Yet experiments like these were instigated by the Meteorological Council and were very much in the service of the Meteorological Office. They were also restricted to meteorology. There were no more experiments into geomagnetism or ether in pursuit of private research agendas of the kind that Balfour Stewart had carried out. In late 1871, the Kew Committee rejected a request by Stewart to have his beloved rotating disk experiments recommenced at Kew. The committee promptly arranged for Robert Beckley to pack up the equipment and take it to Manchester.[106] Also, after 1871 the Kew staff seldom performed experiments on behalf of private individuals, with the help of grants from the Royal Society government grant or BAAS, as had happened in the 1850s and 1860s. Even in rare exceptional cases, such as a photometer sent to Kew by Henry Roscoe, chemistry professor at Owens College in Manchester, this type of work was not given priority. In an 1875 letter to Scott, Roscoe ruefully remarked that "it seems that the Kew Observatory is not the place to get any new method tried . . . ."[107] Kew was no longer a center for private individuals' experiments, as had been envisaged in the original 1842 plan for the observatory.

An increased emphasis on service to other organizations became a characteristic of Kew as a whole after 1871, as is demonstrated by the expansion and development of standardization, the single most dramatic development at Kew between 1871 and 1885. Verifications of the two main classes of instruments hitherto tested at Kew—barometers and meteorological thermometers—only gradually increased up to the mid-1870s. But at the end of the 1860s, Kew Observatory began testing a major new class of instrument quite removed from the meteorological, magnetic, and astronomical sciences practiced there. Scott's 1885 history mentions the verification of 269 clinical thermometers in the period 1869–1870. By the 1872 report of the Kew Committee, the number had increased to 1,395; thereafter, a similar number of clinical thermometers was tested at Kew each year up to the mid-1870s (see table 4.1).[108] No contemporary documentation makes it clear why Kew entered this field. In his autobiography, Kew Committee member Francis Galton claims that his invention (with the help of De La Rue) of an apparatus for testing thermometers quickly and accurately made this mass standardization possible. Yet while it is true that Galton did develop such a machine, in which forty thermometers could be placed in a test chamber at any one time, this did not come into operation at Kew until 1874,[109] by which time clinical thermometers had been undergoing tests in large numbers—sometimes outnumbering meteorological thermometers—at Kew for at least two years. The real reason why Kew Observatory began testing so many clinical thermometers around 1870 seems to have more to do with the fact that at precisely this time, thermometers began to be used in large numbers by the medical profession. Their mass adoption was due partly to their becoming compact enough for doctors to carry on their rounds and also to members of the profession gradually—and sometimes reluctantly—adopting a more scientific approach to medical practices in place of an older philosophy that emphasized a good general education and experienced medical judgment.[110]

In addition, the Kew Committee had a strong financial incentive for testing so many thermometers, as Galton himself admitted in his autobiography.[111] The income from the Gassiot Trust, though substantial enough to keep Kew going, did not amount to much more than £600 per annum. In 1874, for no stated reason, it abruptly dropped to £499 and never rose above this amount until 1885. Thus after 1874, the income from the Gassiot Trust was substantially less than the annual grant of £600 that Kew had regularly received from BAAS in the 1860s. The minutes of the July 1874 Kew Committee meeting clearly suggest that finance was a cause for concern: they record a discussion "on the financial condition of the Observatory" and a request for a quarterly statement of income and expenditure.[112] Although Kew continued to supply many thermometers and barometers to the Meteorological Office and the Admiralty, the majority of the fees received for instrument verifications were from instrument makers. In 1874–1875, for example, the total verification income from the Meteorological Office and Admiralty amounted to £57, while that from "Opticians &c." was over £252. In October 1875, the Kew Committee reported with satisfaction that instrument standardization was showing "a very satisfactory increase in utility."[113] Earlier that year, Scott had asked the Office of Woods and Forests if a better path could be made through the Old Deer Park to Kew Observatory, as "the increase in the operations carried on there has rendered such a measure very desirable" and there had been instances of people being unable to find the building in foggy weather.[114] In 1876, the

Kew Committee decided to open an office at the Meteorological Office's London headquarters to receive instruments for verification at Kew in order to save manufacturers the inconvenience of traveling to the observatory. A notice advertising this new service specifically expressed a wish "to afford the public greater facilities for the verification of instruments at Kew."[115]

All this points toward standardization at Kew becoming an increasingly busy commercial operation, eager to please its customers. The financial incentive to expand instrument verifications was further heightened in 1876, when Kew ceased to be the central observatory of the Meteorological Office and so had its funding cut. Already by 1875, income from standardization was approaching parity with that from the Gassiot Trust: in the year up to October 1875, £456 came from standardization, compared with £499 from Gassiot. Table 4.1 shows that in 1878, the £585 from standardization substantially exceeded the £495 from Gassiot; from 1880 onward, receipts from standardization were always higher than the Gassiot income. In 1885, verifications brought in £727, compared to just £491 from Gassiot. Much of the standardization money came from verifications of clinical thermometers, of which well over 8,000 were being tested each year at Kew by 1885. By the mid-1880s, therefore, standardization was easily the largest single source of the observatory's income.

The importance of this increased commercialization was symbolized in 1877 by the introduction of a distinctive "KO" hallmark, designed by De La Rue, etched on thermometers tested at Kew. This hallmarking of instruments was another idea of Galton's: he first suggested it to the Kew Committee in July 1876, and the following spring he tried etching thermometers himself, finding the process surprisingly easy. The committee approved a specimen of the new mark the following July.[116] Hallmarking was initially done for a fee of three pence, but this charge initially led to a disappointing level of demand for the hallmark. In November 1878, therefore, the committee decided that all thermometers that had passed the verification process should be hallmarked free of charge and that no thermometer would be verified if its maker refused to have it marked. The fact that the committee advertised the service in both *The Lancet* and *Nature* is further evidence of the commercial agenda: the committee was clearly trying to attract the lucrative clinical thermometer market as well as the market for meteorological instruments.[117]

In November 1878, the committee requested that "a female assistant" be engaged to engrave the thermometers with the hallmark. The following month, Whipple reported that he had taken on a Miss H. Clements (first name unknown) for this purpose.[118] Alison Morrison-Low has argued that women played an active role in the nineteenth-century scientific instrument trade, often managing their husbands' businesses, but sometimes working independently as well. Some were assistants in instrument-making businesses, working in various roles that included making the glass tubes for instruments such as thermometers and barometers.[119] It is possible that Miss Clements came from some such background, but the Kew Committee likely hired a woman to engrave the thermometers for the same reason that Elizabeth Beckley was employed to take the solar pictures with the photoheliograph in the 1860s: Miss Clements was cheap labor and could be paid on a piecemeal basis. In November 1880, she was dismissed with a month's notice and a tiny gratuity of £3 3 shillings. Miss Clements's dismissal was not due to any lack of demand but to the continued financial pressures on the observatory, for at the same meeting it was noted that Kew's probable income and expenditure for the coming year would be about equal. Also at this meeting, the committee decided to discontinue receiving instruments at the London office, again for financial reasons. A year later, the advertisements in *The Lancet* (and the *British Medical Journal*) were also discontinued.[120]

Yet the demand to have instruments tested at Kew only continued to rise after 1880. Miss Clements was briefly taken on again in 1883 to help cope with the large number of thermometers being sent to Kew for verification.[121] It would seem that the Kew Committee no longer needed to spend precious money on advertising because this was no longer necessary: the name of Kew Observatory was sufficiently prestigious for the leading instrument makers to automatically send their instruments there to be tested. A comment in an 1880 medical textbook to the effect that the only guarantee of the accuracy of a clinical thermometer was to have it tested at Kew suggests that the Kew thermometer-testing service had by now become indispensable to the medical profession.[122] The Kew Committee even tried to register the Kew monogram as a trademark; it was only stopped by the refusal of the Commissioners of Patents to do this.[123] In any case, it hardly mattered: with the ever-increasing number of instruments being sent to Kew for verification, even after the advertisements ceased, "Kew Observatory" was now effectively a trademark.

After 1876, standardization branched out into further fields. The testing of sextants, begun in the 1860s on the initiative of Francis Galton,

TABLE 4.1. Principal Kew instrument verifications, 1871–1885

	1871–1872	1872–1873	1873–1874	1874–1875	1875–1876	1876–1877
Met. thermometers	1219	782	1471	1238	1410	1428
Clinical thermometers	1395	1233	1255	1439	1560	2281
Deep-sea thermometers	0	0	0	0	0	0
Total thermometers[1]	2661	2096	2780	2761	3130	3863
% clinical	52	59	45	52	50	59
Barometers[2]	124	179	160	214	230	209
Sextants	3	0	1	1	0	3
Income from stdsn. (£)	125	236	253	456	419	456
Total income (£)	2084	1979	2401	2642	2801	2437
Percentage total income	6	12	11	17	15	19
Income from Gassiot	600	608	499	499	498	497

Notes: [1] Total thermometers does not include deep-sea thermometers.
[2] Including aneroids.
- Percentage total income: income from standardization as a percentage of the total income.
Source: Data from KCR, 1872–1885.

remained a relatively minor part of the standardization work at Kew until 1881, when the numbers of sextants tested increased sharply to 25, compared with just 5 the previous year (table 4.1). Sextants were initially tested outdoors in the observatory grounds, but from the late 1860s on they were tested in the basement of the observatory, an environment free from draughts and vibrations, where the magnetic instruments were also housed. The 1881 increase in the numbers tested might have been facilitated by the apparatus being made more robust that year.[124] Sextant testing expanded further in 1883, this time on the initiative of Kew Committee member Lord Rosse, who in the previous year had remarked on how few sextants were being tested at Kew and who expressed a wish to raise interest in the accuracy of "an instrument on which the safety of a ship so intimately depends."[125]

In May 1884, the Kew Committee also began "rating"—that is, testing the accuracy of—watches. The stimulus for this seems to have been external demand as well as the initiative of the committee. As early as 1875, the committee had corresponded with John Hartnup of the Bid-

1877–1878	1878–1879	1879–1880	1880–1881	1881–1882	1882–1883	1883–1884	1884–1885
1435	1286	1487	1704	1518	1165	1225	1825
2032	3405	3638	4217	5365	7255	8726	8238
0	53	22	36	27	51	13	38
3595	4828	5344	6085	7261	8610	10240	10268
57	70	68	69	74	84	85	80
222	196	224	202	183	211	208	256
2	4	5	25	36	55	64	130
585	469	538	595	621	615	759	727
3000	2657	2364	2214	2648	2590	3075	2651
20	18	23	27	23	24	25	27
495	495	496	496	496	493	494	491

ston Observatory near Liverpool, where chronometers were tested for that port's merchant shipowners, about the feasibility of starting a chronometer-rating service at Kew. Nothing came of this for several years because the British Horological Institute's view was that the proposed fee of £1 per chronometer would discourage manufacturers from submitting their instruments for testing. The institute's secretary did suggest, however, that there might well be some demand for rating ordinary watches.[126] Then, in 1880, an anonymous article appeared in the watchmaking trade press expressing the wish that some independent institution such as the Horological Institute would issue test certificates for watches along the lines of those issued in Switzerland and Germany. On reviewing the article, members of the Kew Committee decided against starting such a service at Kew, but late in 1881 they agreed that Whipple should rate a chronometer "as an experiment only." They very likely changed their minds because they saw watch rating as a way of increasing the observatory's revenue: the next meeting began with a gloomy financial report in which the question of staff salaries was deferred. Whipple then presented a report on the experimental chronometer test, where-upon the committee immediately authorized Whipple to begin testing other chronometers for a fee of 7s 6d per timepiece.[127]

In its first years, the watch-rating service was largely restricted to or-dinary watches sold to private individuals rather than marine chronome-ters of the kind being tested at Liverpool (and which had long been tested at Greenwich). This was at least partly because Whipple was not satisfied with the accuracy of Kew's existing clock and transit telescope for de-termining the time standard against which the chronometers would be tested.[128] In 1883, the committee agreed to adopt a system devised by Whipple for rating watches, which included a safe for storing the valu-able watches and a device for testing their accuracy at different tempera-tures. The cost to the Kew Committee was kept to an absolute minimum: for example, the watches were tested against a chronometer set by taking transits of stars at Whipple's own house (with a transit telescope lent by the Royal Astronomical Society) and then taken to the observatory each day. The commercial nature of the work is underlined by the Kew Com-mittee's setting up a system that enabled makers to conveniently deposit their watches at the Meteorological Office or the Horological Institute in central London for onward transmission to Kew. During the twelve months up to November 1885, watch-rating fees brought a further £185 to the observatory—in addition to £727 for other standardization work.[129]

The change of priorities at Kew after Sabine's departure was reflected in the different use of space in the buildings there. For example, a room formerly occupied by Sabine's magnetic clerks was by the early 1880s occupied by assistants employed in standardization work, while another room formerly used for experiments on pendulums (also an interest of Sabine's) had now been modified for testing sextants. Of the fourteen staff employed at Kew in 1885 (figure 4.2), four worked full-time in standardization and a fifth worked part-time.[130] Standardization could be said to have employed even more of the staff at Kew if we include under this heading the testing of magnetic instruments for other observatories.

Instrument standardization maintained and enhanced the interna-tional reputation that Kew had already gained. For example, in 1881 Leonard Waldo of Yale College in the United States discussed a rigor-ous test that he had performed on three Kew thermometers in which he had measured each degree separately. He found that their errors were "practically insensible and too small to be detected with certainty."[131] It was symbolic of the observatory's status that it was no longer possible

FIGURE 4.2. Group photograph of Kew Observatory staff, about 1885. George Whipple, superintendent of Kew Observatory from 1876 to 1893, is seated in the center of the front row. Image courtesy Met Office National Meteorological Library and Archive.

to change its name. The Royal Society Council received such a request in January 1883 from William Thiselton-Dyer, assistant director of the neighboring Royal Botanic Gardens. He asked if the observatory could be known as something other than "Kew Observatory," as confusion with the Botanic Gardens was causing mail to be wrongly delivered. Moreover, the observatory, when built for George III, had originally been known as the King's Observatory at Richmond. Thiselton-Dyer's request for a complete change was firmly rejected by the Royal Society: the only change it agreed to was to call it "The Kew Observatory, Richmond."[132] It must have been hard to imagine either the Kew Committee or the Royal Society dropping the name "Kew Observatory" now that it had become a name and a brand worldwide. The name "Kew Observatory" had stuck.

CONCLUSION: "A PRIVATE ESTABLISHMENT"

In the early 1870s, Kew Committee member William Spottiswoode wrote to Balfour Stewart regarding a difficulty in supplying observational results to Stewart's Manchester colleague Henry Roscoe: "Kew is, as you know an establishment no longer supported by the funds of a public body." Spottiswoode wrote these words after crossing through a version that describes the observatory as "a private establishment not a public one."[133] The crossed-out version is the most revealing—it shows a member of the Kew Committee acknowledging that Kew after the Royal Society's takeover was, in effect, a private institution. Kew Observatory had, of course, been "private" ever since 1842 in the sense that its main source of income was an annual grant from BAAS, which was itself funded by members' subscriptions and donations. Yet it should be clear from this chapter that after the transfer of responsibility to the Royal Society in 1871, Kew Observatory's financial viability was by no means guaranteed and it had to seek its own sources of income. This led to Kew Observatory becoming a very different kind of organization from what it had been under BAAS earlier in the nineteenth century.

Secondary sources are correct in asserting that the motive of the BAAS Council in deciding to relinquish Kew Observatory in 1870 was financial. The annual grant of £600 was too heavy a burden on an organization with limited funds and many competing priorities. It is very likely that Sabine, as president of the Royal Society, used BAAS's decision as an opportunity to gain ultimate control over Kew Observatory. Indeed, he may even have manipulated it to suit this aim: BAAS's decision to contemplate the "dissolution" of Kew and close it down in 1872 gave Sabine the perfect excuse for the Royal Society to take it over. Sabine's subsequent actions substantiate this notion. When the elderly but still influential John Herschel raised objections to the Royal Society assuming responsibility for Kew, Sabine allayed any fears among council members by assuring them that the observatory's new managing committee under the Royal Society would be identical to the existing Meteorological Committee—without pointing out that the Gassiot Trust deed did not make this legally binding. Sabine completed his control over Kew by appointing as superintendent Samuel Jeffery, whose principal qualification for the position seems to have been his loyalty to Sabine.

It is easy to see Airy's failure to transfer the Kew meteorological observations to Greenwich as being due to the machinations of Sabine, his old rival, but it really owes as much to the fact that responsibility for the Kew observations was vested in the Meteorological Office, which was run by the Board of Trade, a separate department of government from the Admiralty. Airy had no authority in the affairs of the Board of Trade. This demonstrates how, by the 1870s, meteorology in Britain had gained official status as a science quite separate from astronomy. Airy's successor as Astronomer Royal from 1881, William Christie, made no attempt to tread on Meteorological Office ground the way Airy had. Edward Maunder's 1900 account of the Royal Observatory modestly describes Greenwich as one of many stations reporting to the Meteorological Office.[134] This official separation of meteorology from astronomy parallels a similar development that happened in France in the 1870s with the establishment of the Bureau Central Météorologique, which ran French meteorology separately from Paris Observatory.[135]

Airy's successful acquisition of the Kew photoheliograph, in addition to his meteorological maneuvers, goes a long way toward explaining why Airy became president of the Royal Society in 1871 and then resigned from this prestigious position after just one year. Becoming president gave Airy the opportunity he wanted to bring the Kew meteorological and solar observations into his Greenwich empire. Even more importantly, it enabled him to undermine the power of the Devonshire Commission's calls for a separate solar observatory, a prospect that was anathema to Airy. This ulterior motive to take the wind out of the commission's sails may be an example of a moderate measure of modernization and diversification being instituted in order to weaken the case for much more radical reform—as Roy MacLeod has suggested was the aim of some in the Royal Society reform movement in the 1840s.[136] Airy's masterful undermining of the proposal for a separate solar observatory shows him deviously manipulating the laissez-faire system by offering to do the same work at Greenwich for less money.

Historians, as well as Robert Scott's 1885 account, have all noted how the Gassiot Trust saved Kew Observatory from closure. This is true, but simplistic. In any event, the Gassiot money proved insufficient to meet the observatory's running costs, especially after the abrupt drop in dividends from the fund in 1874. The reduction in the Meteorological Office grant two years later made matters worse. The smallness of the grants from the Gassiot Trust and the Meteorological Office forced the Kew Committee to change its mission, which it was free to do after Sabine was incapacitated from the mid-1870s. Above all, these financial constraints led to standardization becoming the central feature of the work at Kew and to this standardization work diversifying into fields far beyond geomagnetism and meteorology. Importantly, the majority of these instrument tests were for private instrument makers, not government departments. The opening of a London office for the reception of instruments in 1876, the establishment of a hallmark the following year, and the steep climb in the numbers of clinical thermometers tested from the mid-1870s further point to the observatory becoming more commercially orientated after the drop in grant income in these years. By the time Scott wrote his 1885 paper, Kew was less an experimental observatory of the type conceived in 1842 than a laboratory that tested scientific instruments, predominantly on a commercial basis. Spottiswoode's crossed-out remark of the early 1870s was prescient: Kew was now indeed "a private establishment" that had to earn its keep.

5

Kew Observatory and the Origins of the National Physical Laboratory, 1885–1900

It has been represented to me by Dr. Schuster that at the present juncture it might be well for us to communicate [with] the Kew Committee of the Royal Society, [&?] consider whether it is desirable or feasible, or both, to utilise that Institution as a nucleus of the proposed National Laboratory.

The Chairman of the Kew Committee, Mr. Francis Galton . . . has also been good enough to write similarly. . . .

OLIVER LODGE, 22 FEBRUARY 1893

The present work of the [Kew] Observatory is therefore of a character which is strictly consistent with a large portion of the work which would find a place in a national physical laboratory.

BAAS REPORT, "ON THE ESTABLISHMENT OF
A NATIONAL PHYSICAL LABORATORY," 1896

THROUGHOUT THE LAST FIFTEEN YEARS OF THE NINETEENTH CENTURY, Kew Observatory continued to be run by the Kew Committee of the Royal Society. Then, in 1900, it became part of the newly founded National Physical Laboratory, and the building remained the first site of the NPL until the laboratory moved to its present-day site at Teddington in 1902. The origins of the NPL have been the subject of several scholarly and semipopular accounts, yet none of these describe what actually happened at Kew after 1885, nor whether and to what extent events at Kew influenced—or were influenced by—the NPL. All of them merely note that Kew Observatory offered a convenient first home for the NPL while giving little acknowledgement to any role that the Kew Committee might

165

have played in the NPL's origins. These histories tend to tell the story of the foundation of the NPL as that of an organization intended to rival imperial Germany's flagship national laboratory, the Physikalisch-Technische Reichsanstalt (PTR), built in 1887 at Charlottenburg on the outskirts of Berlin. More importantly, they present the establishment of the government-supported NPL as a triumph of twentieth-century state-supported science over nineteenth-century laissez-faire ideology. Such is the story told by Russell Moseley, Peter Alter, Edward Pyatt and—albeit with slightly more historical context—Eileen Magnello.[1]

This chapter argues that Kew Observatory was much more important to the origins of the NPL than has hitherto been acknowledged. Indeed, I contend that the establishment of the NPL was in many ways a change of name to an existing institution, because well before 1900 Kew Observatory had changed from being a magnetic, meteorological, and experimental observatory into an institution that made most of its income from standardization work. It is clear that this trend was already well under way by 1885. In this chapter, I describe how the change of direction toward standardization became even more marked after this date, so that by the late 1890s Kew Observatory was already doing a substantial part of what Douglas Galton, Oliver Lodge, and others believed should form the program of work for a national physical laboratory. I thus argue that Kew Observatory was a precedent for the NPL and also a center around which it could be built.

This chapter's first section argues that in the 1890s, Kew Observatory became effectively a self-supporting company in its own right, rather like the NPL was expected to be in its first years. In addition, the Kew Committee became dominated by university-based physicists, as the NPL's governing body would be. The following section describes how Kew in these years became primarily a standardization laboratory, with magnetism and meteorology now almost afterthoughts. The final section narrates the story of the NPL's immediate origins in relation to Kew Observatory. From this I argue that Kew Observatory and its committee were central to the planning of the NPL and that the NPL began as Kew Observatory under a different name. In demonstrating this continuity between Kew and the NPL I thus present a challenge to existing narratives of early twentieth-century science as a new departure from the nineteenth-century world of laissez-faire.

THE WORKING OF KEW OBSERVATORY, 1885–1900

The continuing trend toward standardization being the most profitable activity at Kew meant that by the end of the nineteenth century the Gassiot Trust became a small, secondary income. In 1885, the observatory earned £727 from standardization versus £491 from the Gassiot Trust; in 1899, the last year before the NPL takeover, standardization brought in £2,175, compared to just £455 from Gassiot. Even by 1890, standardization was generating more than three times as much money as the Gassiot Trust.[2] The observatory was now effectively self-supporting.

Such was the situation in March 1891, when the Kew Committee asked for the Royal Society Council's permission to apply for a license of incorporation under the 1862 Companies Act. The Meteorological Council, the governing body of the Meteorological Office, applied for such permission at the same time. The Meteorological Council's stated reason for applying was to protect itself "against the possible inconveniences which might at any time arise in connection with their business transactions from the peculiar constitution of their body."[3] The Kew Committee's rationale is not recorded, but the initiative on the Kew Committee was taken by Richard Strachey, then chairman of the Meteorological Council, suggesting that the committee applied for the same reason as the Meteorological Council. Strachey might well have thought that incorporation would make Kew Observatory (and the Meteorological Office) less of a risk from the Royal Society's point of view. The Kew Committee was legally appointed by the Royal Society, yet its ever-growing business with scientific instrument makers and watchmakers also carried with it an element of risk. Manufacturers defaulting on payments could potentially have made the Royal Society liable for large amounts of money. Such debts are likely to have been "the possible inconveniences" resulting from the Meteorological Council's "business transactions," and it is reasonable to suppose that this applied to the Kew Committee as well. Incorporation under the Companies Act would make those serving on the Kew Committee effectively members of a company in its own right and therefore liable for any debts they incurred. This is further evidenced by an amendment to the Kew Committee's proposed memorandum and articles of association under the Companies Act allowing the Royal Society to retain control of the Gassiot Trust while absolving it from responsibility for the Kew Committee's actions.[4]

On 9 February 1893, the Kew Committee officially became registered as the "Incorporated Kew Committee of the Royal Society."[5] Being granted a seal of incorporation was not only a prestige symbol in its own right. More importantly, it meant that the members of the Kew Committee officially acknowledged that Kew Observatory was now primarily a *business*, something it had effectively been for some years. This was symbolized by the incorporation of the Kew Observatory monogram—now famous and a trademark in all but name—into the seal of incorporation. By the mid-1890s, the Kew Observatory business was making enough money to invest its own revenues. At the end of 1894, partly because the bank balance was "unusually large," the committee purchased £900 of India stock. A further £400 of India stock was bought in November 1896.[6] That Kew was now a business, operating practically independently of the Royal Society and mostly supporting itself through fees for testing instruments, set an important precedent for the formation of the NPL, which was initially envisaged by the government as an institution that paid a large part of its running costs from verification fees.[7]

George Whipple did not live to see the Kew Committee's incorporation. He died on 8 February 1893, aged fifty. Since mid-1892 he had been suffering from an undefined serious illness that had prevented him from working, possibly related to the rupture of a blood vessel sustained in 1883. During his illness, his duties were covered by his chief assistant, Thomas Baker—though there is evidence suggesting that Whipple was still signing instrument test certificates within a few weeks of the end of his life.[8] The need to appoint a successor was not unexpected, for in January 1893 Francis Galton had written to the Royal Society Council about the possibility of granting a pension to Whipple after Easter that year, "when his full salary will cease to be paid," indicating that Whipple was already planning to retire due to ill health.[9] After Whipple's death, Baker carried on the day-to-day running of the observatory, but the position of superintendent did not go automatically to him. Baker's name does not even appear on the list of candidates who applied for the position. Instead, when it met on 24 February 1893, the Kew Committee decided to advertise the post in *Nature* and *The Athenaeum* for a period of three weeks. Those answering the advertisement were to be sent a printed leaflet outlining the duties of the post.[10]

The job description leaflet tells us much about the kind of person the Kew Committee was looking for. Its first paragraph is a brief description of the main work at Kew: "The primary work at the Observatory is carrying on meteorological and magnetical observations with self-recording instruments. It also includes experiments in various directions, and the verification of scientific instruments of numerous descriptions." By 1893 the "primary work" at Kew was really verification of instruments, yet this has been given secondary importance here. The description seems to have been slanted in order to attract a particular type of applicant: by presenting Kew as a research institute, the Kew Committee might have hoped to attract a researcher of the highest caliber. If the leaflet had presented the work at Kew as predominantly routine testing of instruments, the committee might have had a less enthusiastic response from the best potential candidates now emerging from the university physics laboratories of the late nineteenth century. That the Kew Committee was looking for a high-flying candidate is further evidenced by the list of qualifications specified in the leaflet: as well as knowledge of the sciences practiced at Kew, the committee was seeking someone with "experimental aptitude, business habits[,] administrative faculty, energy, and scientific status."[11] Placing the phrase "scientific status" at the end of the sentence makes it resound in the reader's mind, suggesting that this might have been an especially important quality being sought.

By 24 March, the Kew Committee had received forty enquiries about the post; by mid-April, twenty individuals had sent in applications. A subcommittee looked at the applications and reported back to the Kew Committee. Of the subcommittee's four members, two (William Grylls Adams and Arthur Rücker) were university physicists, one (Robert Scott) was in a full-time, paid position as head of the Meteorological Office, while the other (Francis Galton) was a self-funded gentleman-scientist.[12] The subcommittee short-listed the applicants to just four, all of whom indeed had "scientific status." Herbert Tomlinson was a demonstrator and lecturer at King's College, London, and had been a Fellow of the Royal Society since 1889. William Dampier Whetham had gained a First in the Cambridge Natural Sciences Tripos in 1889; he was now a researcher in the Cavendish Laboratory and a fellow of Trinity College, Cambridge. Charles Chree also worked at the Cavendish and held a Cambridge fellowship at King's College. Only Thomas Blakesley, an instructor in physics and mathematics at the Royal Naval College in Greenwich since 1885, was not employed at a university—though even he was a Cambridge mathematics Wrangler.[13]

Eight members of the Kew Committee interviewed each of the short-listed candidates on 28 April 1893. Just one member of this interview panel, Galton, was a gentleman-scientist. Otherwise, the panel was made up of Robert Scott, three university physicists (Adams, Rücker, and George Carey Foster) and three scientific servicemen (Captain William Abney, General Richard Strachey, and General James Walker). The panel decided "unanimously" to appoint Charles Chree (figure 5.1) as the new superintendent.[14] The minutes state no reason for the committee's choice, but we can derive some interesting clues from an undated document that lists all twenty applicants. Here, each applicant is ranked with the letters a, b, or c, according to unspecified criteria. All the short-listed candidates are marked "a," except Tomlinson, who is "b or a." But penciled next to the names of the short-listed candidates are figures that correspond exactly to the ages of these four men in 1893. From this, it seems possible that one factor in selecting Chree was that at thirty-two and with a Cambridge Fellowship, he had the right combination of age, level of experience, and "scientific status." Committee members might have considered Tomlinson, aged forty-seven and an FRS, too senior and overqualified. They might have felt the same about Blakesley, forty-five years old and an established instructor at the Royal Naval College. The twenty-seven-year-old Whetham, on the other hand, might have been seen as too young and inexperienced, especially as an essential part of the superintendent's role was now managing a commercially oriented scientific institution.[15]

Chree was certainly not selected for his background in experimental physics and laboratory measurement that had been the chief characteristics of most of his predecessors from John Welsh onward. At the time of his appointment, Chree was primarily a theoretician. At Cambridge he had gained first-class honors in the mathematical Tripos as well as natural sciences. After graduation he did research into electric currents at the Cavendish Laboratory under Joseph John Thomson, but most of his research output by 1893 was in the theory of elasticity. Even throughout his thirty-two-year tenure at Kew, he tended not to carry out laboratory work in person, preferring instead to write papers based on the results of work done by others at Kew and elsewhere. One obituarist humorously recounts how outdated his ideas about laboratory equipment were: once, when he was asked to update his facilities, he presented a list including "'four rings for a retort stand' and a 'nest of four beakers.'"[16] In fact, Chree was rather like Thomson in the sense that Thomson headed an experimental laboratory yet had a primarily mathematical background.[17] Therefore it is likely that in appointing Chree, the members of the Kew

FIGURE 5.1. Charles Chree (1860–1928), superintendent of Kew Observatory, 1893–1925. Photograph by Maull & Fox. Image courtesy the Royal Society.

Committee were looking not for a hands-on, practical physicist like Whipple but rather a laboratory director who was interested in the results of research done by practical workers reporting to him and who therefore would concentrate on directing that research.

Yet most striking of all is how different was the method of Chree's recruitment from that of his predecessors. All the Kew superintendents from Welsh onward had been appointed at least partly through personal patronage. Welsh was selected on the recommendations of William Sykes and Thomas Makdougall Brisbane. Stewart was one of six applicants for the post, but he had been chosen on the grounds that Welsh, before he died, had recommended that Stewart succeed him. Samuel Jeffery probably owed his appointment to Sabine's influence. Finally, Whipple was appointed internally without any competition. But now, in 1893, the job was advertised openly in *The Athenaeum* (on the front page) and *Nature*—two journals widely read by scientists and in university common rooms—and a detailed job description was sent to those answering the advertisement. The most suitable candidates were then short-listed and interviewed by the Kew Committee, completing a narrative very similar to the process of recruiting professional scientists in the twentieth century.

There is no evidence here of any conscious wish to professionalize the position of superintendent at Kew. Rather, this new appointment through open competition reflected the changed attitudes by this time toward appointments to official positions, notably in the civil service. This was also the case at Greenwich: in 1873, to run the recently acquired photoheliograph, Airy reluctantly had to employ Edward Maunder, who had passed the civil service examinations, rather than a handpicked man from Cambridge, as had been his custom.[18] But more particularly for Kew Observatory, the change reflected the extent to which the makeup of the Kew Committee had altered by the early 1890s. Gentlemen-scientists like Gassiot and De La Rue no longer dominated it. Even Galton, though self-funded, firmly believed in science as a full-time occupation; indeed, shortly after Whipple's death, he had written to the Royal Society Council outlining a procedure for appointing a successor, suggesting that it might have been Galton's idea to advertise the post and appoint an outsider rather than make an internal promotion.[19] Of the committee that interviewed Chree and the other candidates, the largest categories represented were the scientific servicemen and university physicists. The servicemen had always been represented on the Kew Committee to some extent, but the university physicists were a new category that had emerged on it since the 1870s. As Romualdas Sviedrys and Graeme Gooday have shown, the last third of the nineteenth century saw a rapid growth of research and teaching laboratories in a number of British universities, stimulated partly by the research required by the electrical industry and also by the need for teaching laboratories to train a new generation of engineers and science teachers.[20] The directors of these laboratories—among whom were Adams, Foster, and Rücker, the three academic members of the Kew Committee—had a strong interest in standardization and precision instruments such as thermometers because of the need to train their students in the all-important skill of measurement. Therefore Kew Observatory, by now world famous for its standardization work as well as magnetic and meteorological measurements, was of great interest to these three.

David Grier has argued that the Board of Visitors that oversaw the work of Greenwich Observatory, and also the committee inspecting the Nautical Almanac Office, was rather like a company's board of directors overseeing the factory production of astronomical observations because they represented "those with a stake in the institution."[21] With the increasing influence of the university physicists, the Kew Committee had now become like a board of directors as well, for the physicists had a similar stake in Kew. In contrast to the middle years of the century, when the likes of Gassiot and De La Rue were personally involved in the experiments at Kew, the committee was increasingly dominated by physicists who were more interested in the *results* of that work, especially standardization. In fact, with its seal of incorporation, the Kew Committee was literally a board of directors, something emphasized by the job description of the superintendent as "the chief executive officer of the Kew Committee, to whom he is responsible . . . ."[22] In appointing Chree, the university physicists were recruiting one of their kind: young, enthusiastic, and with training and background in their community.

Yet some correspondence that took place in the days before the 24 February meeting suggests the most important clue of all as to why Chree was selected. A few days before the meeting, Kew Committee chairman Francis Galton had sounded out Liverpool physics professor Oliver Lodge as to the possibility of extending Kew Observatory in order to turn it into a "national physical laboratory," the need for which Lodge had urged in a speech to BAAS in 1891 (see later in this chapter). Lodge was chairman of a small committee of university physicists set up by BAAS to look into the idea of such a national laboratory. On receiving Galton's letter, Lodge wrote to the other members of this committee, mentioning that the Kew Committee was about to discuss the appointment of a new superintendent and commenting that "it is plain that the possibility of developing the Kew Observatory into an institution comparable with some of those which exist in the capitals of the Continent must depend to some extent on the kind of man they may decide to go for."[23] Another member of the BAAS committee, Richard Glazebrook, deputy director of the Cavendish Laboratory in Cambridge, similarly suggested that the Kew Committee should bear in mind the possibility of developing Kew into a larger laboratory "in any plans they may form for filling up the vacant post."[24] It is likely, therefore, that the Kew Committee appointed Chree as somebody who might one day have to take on the responsibility for a much larger organization than any of his predecessors had done. Chree must have been the candidate most ideally suited to a post with such development potential. He was young—yet old enough to have some managerial experience—and was fresh from the Cavendish Laboratory, which by the early 1890s had earned a reputation as an electrical standards laboratory in addition to being an elite training center in experimental physics.[25] Chree also had sufficient "scientific status" to hold what might become a senior position in British physical science.

Chree began work as superintendent at Kew on 15 May 1893. Baker, who had run the observatory throughout Whipple's illness and after his death, never rose above the position of chief assistant. Baker was paid £250 per annum, though he was given a bonus of £50 each year from 1897 to 1900 inclusive,[26] perhaps in recognition of the extra work involved while negotiations were in progress to develop Kew into the proposed National Physical Laboratory. The salaries of the other senior assistants were capped at £150: when they had reached this amount, they could not be increased further. Occasionally the assistants were paid small bonuses. In February 1893, for example, three were given bonuses of £10 each.[27] The Kew Committee's refusal to increase the salaries caused much frustration among the assistants in the 1890s, for while their salaries remained fixed, the level of responsibility they had to take on greatly increased, especially with the ever-increasing number and variety of instruments being sent to Kew for testing. This is made clear in a letter to the committee from senior assistant E. G. Constable (first name unknown), who by 1893 had worked at Kew for twenty years and was now, he pointed out, in charge of all the watch and chronometer testing: "The work entails much responsibility, and my salary after so *many years* [sic] service is small indeed, and I shall be very grateful for an increase."[28] The situation had not improved by 1898, by which time the idea of siting the NPL at Kew was being publicly aired. In February of that year, Constable and four other senior assistants wrote to the Kew Committee, pointing out that while at Kew the senior assistants received no more than £150 to £175 a year, their equivalents at Greenwich earned £300 to £450 for quite similar work. Constable and his colleagues believed that "the present low standard of salaries is a source of discouragement" and asked if committee members could use their influence to increase salaries if Kew were to become part of the proposed new laboratory.[29]

Yet it was not just the assistants who were poorly paid. Even Chree was paid only £400 per annum (rising to £500 in 1897), compared with the £1,000 per annum earned by the Astronomer Royal at Greenwich. Furthermore, the published job description in 1893 specifically stated that "no provision is made for a pension,"[30] in contrast to the generous pensions paid to the Astronomer Royal and his assistants at Greenwich. This was despite the great responsibility that went with Chree's role: by 1899 he was in charge of eighteen staff and running a central standards laboratory that tested thousands of instruments each year, including many valuable watches and chronometers.

The junior assistants at Kew were paid correspondingly less than the senior staff. In 1890, for example, William Gough, a nineteen-year-old junior assistant, was on a salary of 19 shillings per week, or approximately £50 per year—no more than what Royal Artillery sergeant John Galloway, the very first assistant at Kew under the British Association, had been paid in the 1840s.[31] Small salary increments were sometimes granted if assistants had certificates of good conduct—as in 1899, when £5 was awarded to each of the junior assistants.[32] Much of the basic clerical and calculation work was handled by temporary "boy clerks" aged between fifteen and seventeen who were paid between 10 and 14 shillings per week on appointment, rising to a maximum of 20 shillings (£1) per week. These were not an innovation at Kew in the late nineteenth century: George Whipple had begun his career at Kew aged just fifteen in 1858. But as the volume of work increased, more clerks were needed: six boy clerks were working at Kew by 1899.[33] The boy clerks at Kew can be considered as roughly parallel to the "boy computers" at Greenwich, who did much of the tedious arithmetic involved in reducing raw astronomical data to a usable form. As at Kew, most of the Greenwich computers were of school-leaving age and were not guaranteed permanent work.[34]

At both observatories, the clerks and computers were cheap and expend-able labor—though as with their more senior colleagues, the Greenwich computers were better paid than their equivalents at Kew. The 10 to 20 shillings per week—that is, £26 to £52 per year—earned by the boy clerks at Kew was at or below the lower end of the £40 to £84 per annum earned by the computers at Greenwich.[35] The Kew Committee does not seem to have emulated the short-lived Greenwich experiment, begun in 1891, of employing women as computers. The only women employed in science-related work at Kew before 1900 were Elizabeth Beckley, who worked on solar photography and analysis of the photoheliograph re-sults, and Miss H. Clements, who temporarily engraved the thermome-ters with the Kew Observatory monogram.

Charles Chree's sixteen-page article for the 1897 edition of the *Record of the Royal Society* gives a predictably uncritical picture of the work done in various parts of the Kew buildings on the eve of the NPL era and says little about the staff.[36] But an unpublished description of the observa-tory produced for the benefit of the NPL's recently constituted General Board, whose members inspected the premises in October 1899, reveals a further detail. The different ranks of staff at Kew were identified by the wearing of colored ribbons: the boy clerks wore blue ribbons, the junior assistants white, and the senior assistants red.[37] Further evidence of the regimented regime at Kew is provided in a later reminiscence, citing sto-ries told by staff who had worked at Kew in the late nineteenth century—including one that claimed that the superintendent stood on the front steps of the building and blew a whistle to call the staff back to work after lunch.[38] Some historians have likened Airy's regime at Greenwich to a factory, in which "profit was measured in terms of public utility and sci-entific prestige, rather than Pounds Sterling,"[39] and it is clear that in this regard Kew bore strong similarities to Greenwich. Right up to the end of the nineteenth century, Kew Observatory remained as hierarchical as it had been in the 1850s, with each man—including the superintendent—in his place. When it became part of the NPL in 1900, it was still a tradi-tional Victorian observatory.

GEOMAGNETISM, METEOROLOGY, AND STANDARDIZATION AT KEW TO 1900

In George Whipple's last years as superintendent, geomagnetic work at Kew continued as it had since Sabine's withdrawal from the scene in the mid-1870s. The magnetic instruments in the basement continued their automatic recording of changes in Earth's magnetic field, and Kew maintained its international status as a magnetic observatory. Kew also remained a center to which overseas observatories sent their instruments to be calibrated: in 1890, instruments for observatories in Hong Kong, Italy, and Portugal were examined at Kew.[40] The observatory was still used as a reference station for magnetic surveys. For example, in 1887 Arthur Rücker and Thomas Thorpe, professor of chemistry and a colleague of Rücker's at South Kensington, visited Kew to make base observations for a magnetic survey of Ireland, part of a new magnetic survey of the whole British Isles. In their resulting paper they used magnetic data from Kew going back to 1858.[41] However, although Arthur Rücker became a member of the Kew Committee in 1889, he ran the magnetic survey in his capacity as professor at South Kensington and not under the auspices of Kew Observatory. Whipple was never pressed into service to make the field observations in Scotland the way John Welsh was in 1857. By the late 1880s, Kew had long since ceased to be the supreme coordinating center for magnetic observations that it had been in Sabine's time.

Geomagnetism at Kew underwent something of a revival after Charles Chree arrived as superintendent in 1893. Although Chree had originally made his name as a physicist in the theory of elasticity, he switched to terrestrial magnetism at Kew. This remained Chree's main research interest for the rest of his career and is the work for which he is best remembered. In 1895 he announced his first major result: that so-called "quiet days," which had hitherto been considered days in which there was little or no magnetic activity, showed magnetic variations. These variations had to be taken into account in studies of magnetic cycles. Chree later extended his research into solar-terrestrial relations. By the end of his career—he only retired from Kew in 1925 and died in 1928—he had greatly strengthened the link already known to exist between solar activity and magnetic disturbances on Earth by demonstrating that such disturbances tended to repeat themselves every twenty-seven days, exactly the same as the sun's rotation period as seen from Earth.[42]

This confirmation of the link between the sun and terrestrial magnetism might seem like Kew Observatory coming round full circle and again becoming the kind of organization that it had been under Sabine. It was Sabine, after all, who had discovered the correlation between the sunspot and magnetic cycles in 1852 and who had transformed Kew into a world-class geomagnetic and solar observatory. Yet there was now an important difference. Although Chree used Kew data in his analyses, he did not work in the capacity of director of a network of observatories across the globe, as Sabine had done. Even the "quiet days" so important to Chree's discoveries were not selected by Chree himself. From 1891, the annual reports of the Kew Committee regularly listed the days "selected by the Astronomer Royal, as suitable for the determination of the magnetic diurnal variations" because they were magnetically quiet.[43] Greenwich and not Kew was now setting the pace as to which days were suitable for these observations. George Airy had retired as Astronomer Royal in 1881. He was succeeded by William Christie, the son of Samuel Hunter Christie, one of those who had attempted to set up a Royal Society magnetic and meteorological observatory in 1840. Christie was by all accounts a more genial figure than Airy, but he still saw himself as the head of a national observatory and so would have believed that direction of observations in solar-terrestrial relationships was his prerogative. Indeed, Christie wanted to expand the "physical" side of the work at Greenwich, including the magnetic and solar observations, which was part of the reasoning behind his new "physical observatory" building at Greenwich.[44]

Although he was a theoretician and not a laboratory worker, Chree did not isolate himself from the day-to-day magnetic work at Kew. For example, an obituarist credits him with personally training magnetic observers for the "Southern Cross" Antarctic expedition in 1899.[45] But one might question whether he needed to have been at Kew to make the discoveries in sun-Earth interactions that formed his most important work. He did the work largely on his own, not as the head of a team at Kew. Also, the sophisticated statistical analyses in his magnetic papers were facilitated by his background in Cambridge mathematics. Chree—a lifelong bachelor—might well have been able to do exactly the same work if he had remained at Cambridge.

Geomagnetism provided a stimulus for the work at Kew to expand into a further observatory science early on in Chree's time as superintendent. In June 1896, Chree received a letter from John Milne, a gentleman-scientist who had become well-known for his investigations of earthquakes. Milne asked Chree if he could make the Kew Committee aware of "the desirability of comparing magnetic & other observations with the unfelt movements of the ground." Milne also suggested that there might be a correlation between movements of Earth's crust and the daily changes of magnetic declination.[46] Milne was one of a number of British academics who had taken teaching posts at Japanese universities in the 1870s in response to rapidly industrializing Japan's drive to train its engineers in Western science and technology.[47] While working as a professor of geology and mining at Tokyo, Milne began a serious study of Japan's many earthquakes and started researching the possibility of recording large earthquakes from anywhere on Earth. By the time he returned to England in 1895, he was a well-known figure in the young but growing field of seismology. He settled in the village of Shide, near Newport on the Isle of Wight, where he established what would quickly become a world-famous seismological observatory. Milne began establishing a system of seismological stations across the globe that aimed to determine the epicenters of earthquakes, using earthquakes to probe the internal structure of Earth and to correlate them with other terrestrial phenomena such as geomagnetism.[48]

Having read Milne's letters to Chree, on 23 October 1896 the Kew Committee resolved to invite Milne to Kew so that he could assess whether it would be suited to seismological observations. That the committee took up Milne's suggestion so enthusiastically was likely related to a vacancy on the committee that year being filled by John Perry, a close friend of Milne's who had worked with him in Japan. Now a professor at the Royal College of Science in South Kensington, Perry attended the 23 October meeting. Milne duly visited Kew on 8 December and recommended a suitable position in the basement of the observatory for setting up his recording equipment. The following year, the Kew Committee successfully applied for £60 from the Royal Society government grant to buy and set up a seismograph of Milne's design, a complex instrument that recorded earthquakes and tremors photographically. It was made commercially by the London instrument maker Robert Munro. Recording of earthquakes began at Kew in 1898.[49] Like the magnetic observations, these seismological recordings were not directed from Kew, which was but one station in a global system of observatories sending its results to Shide.

Although Kew Observatory had in 1876 ceased to be the "central observatory" to which data from the Meteorological Office's network of outlying observatories were sent, throughout the period from 1885 to 1900 senior staff at Kew still had responsibility for inspecting these observatories on behalf of the Meteorological Council to ensure that their instruments were working as efficiently as the Kew standards. Whipple personally inspected some of the observatories; the others were visited by Thomas Baker, the chief assistant. The volume of work increased toward the end of the nineteenth century as the Meteorological Council added several new stations to its network. By the late 1890s, in addition to the original six established in 1867 (Aberdeen, Glasgow, Stonyhurst, Armagh, Falmouth, and Valencia), there were now also stations at Fleetwood, Deerness (on the Orkney Islands), Fort William, Alnwick Castle, North Shields, Yarmouth, and Dublin.[50] In 1890 the Meteorological Council also established an observatory at the summit of Ben Nevis, the highest point in Britain, with an accompanying outstation at the base of the mountain. In the summer of that year, Baker duly visited the observatory at the base to set up the barograph and thermograph.[51]

The experiments on cloud photography, begun in 1885, continued at Kew until 1892. In 1888 these were extended to time-lapse photographs of cirrus clouds, taken with one camera, which were successful in showing extensive structural alterations in these high clouds at time intervals of as little as two minutes. As before, the work was not independent but directed by the Meteorological Council.[52] By the early 1890s, Kew was by no means the only observatory photographing clouds. In 1891, BAAS set up a committee to coordinate a national project to photograph and classify cloud formations, and Kew contributed some photographs at this committee's request.[53] This was an ironic reversal of roles that had been in place since the middle of the century, when the Kew Committee, then under BAAS, had coordinated observing efforts such as the ground observations made by volunteers during John Welsh's balloon ascents in 1852. Now, the Kew Committee of the Royal Society was humbly complying with a request from BAAS. Similarly, in 1894 Kew contributed cloud photographs to an exhibition organized by the Royal Meteorological Society. As early as 1887, this society had sent a circular to some two hundred private individuals with an interest in meteorology, asking them to take photographs in order to resolve queries as to the nature of lightning and cloud formations. As Jennifer Tucker has described, simple handheld cameras became widely available during the 1880s, as did "dry" photographic plates that required no complex preparation before exposure and were easy to develop afterward. There was therefore a large pool of volunteer photographic talent, and the RMS hoped that the many resulting photographs of clouds would have more scientific authority than the often sketchy and sometimes contradictory reports based on visual observations. Thus at the 1894 exhibition, the Kew photographs of clouds were only part of a much larger display in which many of the pictures were taken by volunteers.[54] It was a very different scenario from the 1855 international exposition at Paris in which Kew had put on its own display of magnetic and meteorological instruments.

Chree played little direct part in the day-to-day meteorological work after he became superintendent in 1893. From the beginning, he was less involved in the inspections of outlying observatories than Whipple had been. In 1894, for example, he visited the stations at Aberdeen (his native city) and Glasgow, but by the late 1890s he was leaving the work to his senior assistants.[55] One aspect of meteorological research that did show a revival in Chree's early years was atmospheric electricity. This had formed an important part of the earliest observations carried out at Kew after BAAS took over the building in the 1840s. Even after the retirement of Francis Ronalds in 1852 and the subsequent replacement of Ronalds's electrometer in the dome with the photoheliograph, studies of atmospheric electricity never wholly died out at Kew. In 1860, an electrometer designed by William Thomson began working at the observatory.[56] At the 1881 annual meeting of BAAS, Whipple presented a summary of the results obtained at Kew with a Thomson electrometer over the year 1880, together with a discussion on what light they shed on relations between atmospheric electricity and meteorological phenomena.[57]

Soon after Chree's arrival, the electrometer was overhauled and a new series of observations commenced.[58] Chree published the results in an extensive paper in the *Proceedings of the Royal Society* in 1896. Again, however, Chree did not make the observations himself. At the end of his paper, he acknowledged that "Mr. Constable [one of the senior assistants at Kew] took all the electrical observations and the measurements of the meteorological curves."[59] Observations with the electrometer continued after 1896, but in 1899 the Meteorological Council allowed the Kew Committee to lend the results to a near contemporary of Chree's and a fellow graduate of the Cambridge Natural Sciences Tripos, Charles Thomson Rees Wilson. Wilson is now best remembered for the invention of the cloud chamber, which facilitated the study of subatomic particles and enabled Joseph John Thomson to measure the charge on the electron, but his initial interest was meteorology. It is likely that Chree, wanting to concentrate on terrestrial magnetism and otherwise busy with running the observatory, was glad to hand the electrometer work over to Wilson, his old Cambridge colleague. In Wilson's work we see another example of how the Kew Committee was no longer directing the cutting edge of research. Although Wilson used data from Kew, he carried out the research at Cambridge.[60] In meteorology as much as magnetism, Kew continued as a routine monitoring station, yet other bodies—including BAAS and

TABLE 5.1. Principal Kew instrument verifications, 1886–1900

	1886	1887	1888	1889	1890	1891
Met. thermometers	1320	1370	1074	1910	4901	2289
Clinical therm.	9054	8668	10442	10116	12536	15692
Deep-sea thermometers	32	35	77	100	40	58
Total thermometers	11490	12989	13502	12805	18125	18600
Percentage clinical	79	67	77	79	69	84
Barometers[1]	247	202	279	232	268	279
Sextants	139	145	157	292	346	428
Telescopes	0	0	0	99	152	374
Binoculars	0	0	0	341	336	470
Camera lenses	0	0	0	0	0	19
Watches	490	510	639	528	513	709
Marine chrons.[2]	17	14	12	10	3	18
Income from stn. (£)	1072	1229	1422	1327	1597	1629
Total income (£)	3097	3621	3790	3546	3467	3912
Percent total income	35	34	38	37	46	42
Income from Gassiot	490	490	493	488	488	488

Notes: * 30 of these chronometers were tested for the Italian government.
[1] Including aneroids.
[2] Not including marine chronometers for the Portuguese government, 1894.
- Total thermometers does not include deep-sea thermometers.
- Income from verifications includes that from rating of watches and chronometers.
- Percentage total income: income from standardization as a percentage of the total income.
Source: Data from KCR, 1886–1899, and "Report on the Observatory Department, 1900."

the Royal Meteorological Society, as well as Cambridge—were leading the research program.

While Kew Observatory after 1885 no longer played a leading role in geomagnetic and meteorological research, it became more than ever a leading center for standardization (see table 5.1). Tests of clinical thermometers continued to provide a substantial part of the large income earned from standardization, and the numbers tested continued to rise after 1885. The 1886 report of the Kew Committee claims that 9,054

1892	1893	1894	1895	1896	1897	1898	1899	1900
1875	2246	3225	2647	4098	2874	3296	2892	2786
16850	14682	15593	16699	13772	17270	17962	16020	20476
31	69	35	125	74	119	79	19	83
19018	17260	19095	19767	18170	20523	21603	19244	23547
89	85	82	84	76	84	83	83	87
227	237	245	494	309	375	404	367	416
463	517	461	532	591	694	750	876	813
487	913	249	456	546	707	681	561	1345
168	466	417	376	455	661	374	404	963
18	31	27	14	14	10	13	160	136
1044	1521	737	746	583	680	483	469	403
19	40*	21	58	61	65	70	56	53
1857	1865	1944	2004	1996	2309	2196	2176	2550
3848	3485	4173	4377	3507	3848	4226	4394	8294
48	54	47	46	57	60	52	50	31
488	486	484	485	445	459	458	456	454

clinical thermometers were tested at Kew; the corresponding figure for 1899 was over 16,000.[61] But the rise in income must also be due to the great diversification in the range of instruments tested after 1885. Tests of sextants, never more than 100 per year before 1885, rose sharply in 1889, and from the mid-1890s more than 500 were tested every year. Also in 1889, Kew began testing large numbers of telescopes and binoculars for the Royal Navy.[62] As with sextants, around 500 telescopes and a similar number of binoculars were being tested at Kew each year by the late 1890s. In 1891, Kew also began testing camera lenses, though until the late 1890s never more than 31 lenses were tested each year. Only in 1899 did the figure jump to 160 lenses.[63]

Kew Observatory's annual allowance from the Meteorological Council remained static at £400 throughout the last fifteen years of the nineteenth century, while the income from the Gassiot Trust continued to shrink. The Kew Committee therefore had a clear financial incentive for diversifying the standardization work. Yet there is also evidence that Kew began testing some new types of instruments in response to customer demand. The testing of Royal Navy telescopes and binoculars was first proposed in 1888 by the secretary of the Admiralty, who asked that all such telescopes be tested at Kew before purchase. The following year, the Kew Committee began hallmarking these naval instruments with the "KO" monogram that had become famous on the thermometers—a sign that the Admiralty wanted the prestige of instruments bearing the Kew hallmark.[64] The navy likely needed to purchase a large number of optical instruments as part of its initiative in the late 1880s to build a new generation of battleships of the latest design.[65] In 1886, George Whipple was strongly rebuked by an agent of the Admiralty for refusing to test some range finders for the navy. The agent offered to carry out some experimental testing at Kew at his own expense, and strengthened his request by appealing to Whipple and the Kew Committee's sense of patriotic duty: "I hope the Committee will not refuse this favor [sic], especially considering that it is a work done for the Admiralty, and with instruments to be used on board the ships of the Royal Navy. . . ." The Kew Committee must have complied, for its report the following year records that seventeen range finders were tested at Kew.[66]

The Kew Committee actually had to reject a proposal to test the lamps, lenses, and shades on ships' lights for the Merchant Navy on account of the observatory's inland position, far from any port, in addition to the lights being heavy and bulky to transport and there being no money to set up a suitable outstation at a port.[67] However, the usefulness of Kew Observatory's instrument-testing service to the navy and merchant marine was made clear when this proposal's instigator, Kew Committee member Captain Ettrick Creak, remarked, "I hardly can doubt the full success of the project of testing lights or at any rate lenses and burners if we had a 'Kew' at the East End." Creak was yearning for a "Kew" in the East End of London, the location of the British capital's dockyards and, more particularly, the Admiralty Compass Department where Creak worked.[68]

External demand was behind another striking innovation at Kew after 1885. Hitherto, the observatory's watch-rating service had been mostly restricted to ordinary watches, but in 1886 Kew began testing, on a commercial basis, marine chronometers for determining longitude at sea. In November 1885 the Kew Committee received a request from Dent, the well-known clock maker, to have chronometers tested at Kew. The committee lost no time in acquiring a secondhand oven from the Meteorological Office to test the timepieces' performance at a variety of temperatures, representative of the different climates encountered on ocean voyages.[69] Kew began testing chronometers in August 1886. The trial initially used at Kew tested the chronometers for five periods, each lasting six days, at temperatures between 55 and 80 degrees Fahrenheit. The chronometers were given one day's rest between each temperature test, resulting in an entire trial lasting thirty-five days. If the difference between the chronometer's daily error (or "rate") at the different temperatures was sufficiently small, it was given a certificate.[70] A total of seventeen chronometers were tested at Kew in 1886. The number declined over the next few years, before gradually rising again and peaking at seventy in 1898.[71]

Chronometer rating was Kew's biggest incursion yet into territory traditionally held by Greenwich. Indeed, it formed part of Greenwich's original mission, going back to 1675, of enabling sailors to find longitude. In Airy's time, for Kew to be testing the most important piece of equipment for finding longitude would have been unthinkable. However, Astronomer Royal William Christie's desire to expand the work at Greenwich beyond its traditional navigational remit, together with Christie's relatively easygoing personality, might explain why Christie was tolerant toward Kew Observatory rating chronometers. Indeed, Christie was initially very helpful in his attitude to chronometer rating at Kew. In 1888, responding to a letter from Whipple, Christie recommended a more stringent test than that in current use at Kew. He advised that Kew should increase the maximum temperature used in the test to 95 degrees Fahrenheit and also reduce the allowance it made for changes in temperature when determining a chronometer's rate.[72] The Kew Committee eventually implemented these recommendations. It called this tougher test the "class A" test, while a modified version of the old test was called "class B" and charged an accordingly lower price.[73]

However, in the early 1890s, with the introduction at Kew of a chronometer test as tough as that at Greenwich, Christie's staff at Greenwich did come to view the chronometer-rating business at Kew as competition. In May 1893 Thomas Lewis, an assistant astronomer at Greenwich, warned an official at the Admiralty about an advertisement for chronometer tests at Kew and of the need to take immediate action "to stop the Clerkenwel [sic] people sending their best chronometers to Kew." Lewis suggested that "if a possible number to be purchased could be named we might do this." Clerkenwell had long been an instrument-making district in London. The official replied that the Admiralty could not "interfere" with the Kew advertisement, but reassured Lewis that he could tell the chronometer makers that the Admiralty would be willing to spend around £1,000 on chronometers that year, "provided they are recommended for purchase by the Astronomer Royal." Ten days later, Lewis had written "to the Clerkenwell people."[74]

That Greenwich was thus tipped off about the Admiralty's plans to purchase a large number of chronometers might explain why only ten chronometers were tested for British makers at Kew in 1893. But as the 1890s progressed, the number of chronometers tested at Kew continued to rise: in the long run, Greenwich was unable to stop makers from having their instruments tested at Kew. In any case, Kew may have been no great threat to Greenwich, whose principal client was the Admiralty. The Kew Committee could sell its services to other customers, not least the merchant shipping sector, which it targeted in 1891 with a circular sent to all the main British steamship companies, advertising chronometer rating at Kew.[75] More dramatically, in April 1893 the committee agreed to a request from the Italian naval attaché in London to test some chronometers being bought by the Italian government. Beginning in June 1893, some thirty chronometers were tested in a specially built oven in the basement of the observatory at temperatures of up to 100 degrees Fahrenheit, presumably to allow for the Italian climate. The following year, the Kew Committee performed a similar trial of chronometers for the Portuguese government.[76] Both requests indicate that Kew now had an international reputation in this field, in addition to its established name in testing magnetic and meteorological instruments.

By the 1890s, this renown was also firmly established among makers of ordinary watches, whose letterheads proudly claimed that their wares were tested at Kew and Greenwich. Makers could choose to have watches tested at Kew in one of three classes of trial: A, B, and C, of which class A was the toughest. The performance of watches in class A was further assessed with marks between 0 and 100, with a watch only just passing the class A standard being given 0 and a watch with no measurable error at all assigned 100. The cream of the class A watches were given certificates marked "especially good."[77] The watches scoring the highest marks in each year's tests were listed, along with their marks and the manufacturers' names, as an appendix to each year's report of the Kew Committee, which turned the Kew class A test into a fierce competition among the elite watchmakers of London and Coventry. By the late 1890s, some chronometer makers were calling for the marks to be made less explicit because they found that watches with even slightly lower marks than their very best ones were taking longer to sell.[78] That one manufacturer in 1899 allegedly exaggerated the number of watches he had had tested at Kew is further evidence of the importance of Kew Observatory to the British watch trade in the 1890s.[79]

As with geomagnetism and meteorology, from the mid-1890s standardization at Kew became influenced by the Cavendish Laboratory. Hitherto, thermometers tested at Kew had generally measured temperatures of up to 100 degrees Celsius. In January 1895 the Kew Committee discussed obtaining a "platinum thermometer," a new type of thermometer that measured temperature using the resistance of a length of platinum wire.[80] Platinum thermometers could measure much higher temperatures, up to around 600 degrees Celsius, so they had many industrial as well as scientific uses, such as measuring the freezing points of metals. Also, they could determine temperatures far more precisely and could be used in standardizing conventional mercury thermometers.[81] In October 1895 the Kew Committee obtained £100 from the Royal Society government grant to begin investigations into the working of platinum thermometers.[82] Three months later, Charles Thomas Heycock of the Cavendish Laboratory visited Kew to compare the performances of platinum and mercury thermometers at temperatures above 100 degrees Celsius. In 1896, the Manchester physicist John Allen Harker, who had recent experience in working with platinum thermometers, also offered his services to the Kew Committee. In particular, Harker proposed to compare the performance of platinum thermometers with a gas thermometer used by the Bureau International des Poids et Mesures (BIPM) at Sèvres, near Paris.[83] The BIPM had been opened in 1878 as an international standardization center for metric weights and measures authorized by an 1875 treaty under which seventeen countries—excluding, controversially, the United Kingdom—committed themselves to the metric system.[84] Harker's collaborative work with the BIPM was published in an 1899 paper by Harker and a physicist at Sèvres, Pierre Chappuis. In November 1899, just weeks before the handover to the NPL, the Kew Committee authorized the building of an apparatus for comparing mercury thermometers with a platinum thermometer.[85]

Harker was not a Cambridge graduate: he had studied at Owens College in Manchester and then obtained a PhD in Germany before returning to Manchester to work with Arthur Schuster, Balfour Stewart's successor at Owens College. Yet it is noticeable that William Napier Shaw, another Cavendish physicist who had been on the Kew Committee since 1894, helped with the platinum thermometry project—for example, in deciding on a suitable room in the Kew buildings for the experiments to take place. Technical assistance was lent by the Cambridge Scientific Instrument Company, of which Shaw was a director and that had begun making platinum thermometers commercially.[86] A remark in the 1897 Kew Committee report indicates that the committee and the staff at Kew were under some pressure to produce results in this new field: "This is a subject of increasing urgency in view of repeated requests for direct high temperature verifications which cannot as yet be satisfactorily dealt with."[87] In fact, even before the £100 grant had come in, the Kew Committee authorized Rücker and Shaw to buy the platinum thermometry equipment. It is quite possible that Shaw and the Cambridge Scientific Instrument Company were the source of at least some of the pressure, as the company, which worked closely with the Cavendish and was keen to sell its wares to laboratories, would have had an interest in developing a reliable platinum thermometer with the Kew hallmark as a seal of quality. Indeed, the company's 1893 catalog proudly informed potential buyers of its wares that its platinum thermometers would all come with tables of corrections similar to those that Kew Observatory included with instruments tested there—a further indication of the prestige of the Kew instrument tests.[88] The importance of Harker's work is emphasized by his being taken on at Kew as a "special assistant to the Superintendent" in 1898. He was still listed as such on the payroll after Kew was handed over to the NPL in 1900 and went on to become head of the NPL's heat and thermometric division, where he continued the work he had commenced at Kew.[89] Harker's work on platinum thermometers demonstrates that something like the NPL's thermometric division was already in existence at Kew several years before the NPL itself.

The multiplicity of work being carried out at Kew Observatory by the late 1890s must have made the place seem crowded. The observatory had been extended somewhat since BAAS acquired it in 1842. A one-story outbuilding, just south of the main building, was built in the late 1860s after Kew became the Meteorological Office's central observatory. By 1897 it was known as the "clinical house" because by then it was used for testing the clinical thermometers. In 1887, the Kew Committee agreed that to install additional equipment in the observatory more space would be needed. Accordingly, an additional story was built on top of the east wing of the observatory in 1888, followed by an identical story on the west wing in 1891. The building of both these extensions was financed by loans from the Royal Society Council. The completion of the western extension gave the building the general appearance that it retains today. Yet Chree's 1897 article in the *Record of the Royal Society* glosses lightly over the fact that the prestigious rating of watches was taking place in the same western room on the ground floor where the standard barometers and thermometers, as well as the engraving machine used for making new standard thermometers, had been in use since the 1850s.[90] This is, however, apparent from the document drawn up for the inspection of the premises by the NPL Executive Committee in October 1899.[91] It may be that by the time the NPL took over, Kew Observatory had reached the limits of its spatial capacity. George III's former private observatory had served well as the experimental observatory conceived by BAAS, yet for a national scientific institution branching out into ever more types of work, larger premises were needed.

KEW OBSERVATORY AND THE ORIGINS OF THE NATIONAL PHYSICAL LABORATORY

Standard histories of the National Physical Laboratory and early twentieth-century science typically trace the origins of the NPL to Alexander Strange's 1868 paper on state funding for science and the subsequent Devonshire Commission. Among the recommendations of the latter were physical laboratories for determining national standards.[92] In his 1871 presidential address to BAAS, William Thomson also argued for the setting up of government-supported laboratories that would do research in the physical sciences. Thomson alleged that university physics laboratories had no time for research because they were fully occupied with teaching.[93] But neither Strange nor Thomson called for a single central laboratory supported by the government. The earliest public call for this type of institution came in 1891 from Liverpool physics professor Oliver Lodge. Like the Kew Committee members George Carey

Foster and Arthur Rücker, Lodge was one of the younger generation of professional physicists who held full-time academic posts. On 20 August 1891, Lodge gave a speech to section A (Mathematical and Physical Science) of BAAS in which he urged that physics, especially "the quantitative portion" dealing with standardization and constants, needed to be pursued "in a permanent and publicly-supported physical laboratory on a large scale." Lodge acknowledged the importance of the work currently being done at Kew. He also acknowledged the Board of Trade's standards department at Westminster, which looked after the national standards of weight and measure; since 1889 it had also maintained an electrical standards laboratory that ran a limited program of electrical instrument testing. Now, however, Lodge wanted to see a much bigger establishment on a national scale, "a Physical Observatory, in fact, precisely comparable to the Greenwich Observatory." Such a national institution, Lodge said, would house Britain's national standards and, in an echo of Thomson's 1871 speech, would also carry out long-term researches, for which university laboratories were inadequate because "in most college laboratories, under conditions of migration, interregnum, and a new regime, continuity of investigation is hopeless."[94]

After Lodge's speech, BAAS appointed a committee to discuss the question of such a national laboratory.[95] The committee was chaired by Lodge, and all its other members were—or had been—leading university physicists: Richard Glazebrook of the Cavendish Laboratory, William Thomson, Lord Rayleigh (former director of the Cavendish Laboratory), J. J. Thomson, Arthur Rücker, Robert Bellamy Clifton (Oxford), George Fitzgerald (Trinity College, Dublin), George Carey Foster, and Viriamu Jones (University College, Swansea). Two of these, Foster and Rücker, were serving members of the Kew Committee. Historians have claimed that the BAAS committee was very pessimistic about the prospects of setting up a national laboratory. They cite, for example, Glazebrook as saying that "it was felt to be hopeless to approach the Government" for funds and the fact that the committee's discussions went nowhere. But this seems to be based on reminiscences after the event—in the case of Glazebrook's alleged "hopeless" feeling, an article in *Nature* from 1901 after the NPL had been established, together with Glazebrook's own reminiscences in 1933.[96] In fact, unpublished archival evidence shows that the BAAS committee was still very much alive in early 1893 and that several of its most influential members were far from pessimistic. On 21 February 1893, Lodge, the committee's chairman, replied to a letter from Francis Galton, chairman of the Kew Committee, which apparently suggested that Kew Observatory could form a center around which the proposed National Physical Laboratory might be built. Lodge claimed that Arthur Schuster of Owens College, Manchester, had written to him the day before with a similar suggestion.[97] Whether Galton or Schuster had the idea first we do not know, but it was Galton who seems to have triggered what happened next.

Lodge immediately wrote to the various members of the BAAS committee about the possibility of using Kew "as a nucleus of the proposed National Laboratory" and asking them to send their views urgently to Galton in time for the next meeting of the Kew Committee, scheduled for the following Friday, 24 February. It was in this letter that Lodge suggested that the possibility of developing Kew into a national laboratory would depend on who the Kew Committee would appoint to succeed George Whipple as superintendent.[98] In a separate note to Galton, Lodge asserted that the need to appoint a new superintendent presented "a fitting opportunity" to move forward the idea of a national laboratory.[99] Thus Galton and Lodge between them used the vacancy at Kew as an opportunity to advance the idea of making Kew the "nucleus" of a larger laboratory.

Schuster was the first to respond to Lodge's circular letter. Although he thought that there was "little hope" that the government would right now do anything so radical as fund a new laboratory to compare with the Berlin PTR, he believed that the BAAS committee could certainly obtain part of what it wanted "if the work of the Kew Observatory could be extended in certain directions." Schuster underlined Lodge's view that now was an ideal opportunity to consider whether to extend Kew into a national laboratory. If the Kew Committee did not act now, he said, the movement for a national laboratory might not go anywhere for some time. Similarly, J. J. Thomson had "no hesitation" in asserting that the prospects of founding and running a national laboratory were better under the Kew Committee than under any other organization he knew. Richard Glazebrook, Thomson's deputy, agreed that the activities at Kew Observatory "might form a nucleus around which other investigations might centre."[100] Thus the BAAS committee members who wrote to Galton were certainly not pessimistic about the idea of enlarging Kew to form a national laboratory.

The minutes of the 24 February Kew Committee meeting merely record that Galton had communicated with members of the BAAS committee as to whether they had taken into consideration the existing laboratory facilities at Kew. They also note the encouraging replies from Lodge and the others.[101] No actions or agreement on this subject are recorded in the minutes of this or any subsequent meetings of the Kew Committee before the mid-1890s. The next recorded discussion of the idea of a national physical laboratory did not take place until more than two years later. At the BAAS annual meeting in September 1895, Francis Galton's cousin, Douglas Galton, addressed BAAS as president. His speech included a survey of the assistance currently given to British scientific research in which he noted that Kew Observatory carried out, on a small scale, part of the work done by the PTR in Berlin, but that its further development was "fettered by want of funds." He now suggested "there could scarcely be a more advantageous addition to the assistance which Government now gives to science than for it to allot a substantial sum to the extension of the Kew Observatory, in order to develop it on the model of the Reichsanstalt."[102]

It is important to note that in this speech, Douglas Galton was very conservative in his views concerning state support of science. Earlier in the speech, he said that while the expansion of scientific research meant that scientists "occasionally" had to seek government help, "It would be unfortunate if by any change [sic] voluntary effort were fettered by State control."[103] Galton still supported the laissez-faire system. Even if he was just saying this to please his audience, his use of such rhetoric only emphasizes the ideological environment in which he was speaking. At the same BAAS meeting, Douglas Galton also presented a paper to section A (Mathematical and Physical Science) about the PTR, in which he opined that Kew Observatory "appears to afford a nucleus which might be gradually extended into an establishment analogous to the Reichsanstalt"—using the same language as that used by Oliver Lodge and Richard Glazebrook in their replies to Francis Galton in early 1893. In this speech, Douglas Galton urged that the BAAS committee on the new laboratory be reconvened and report back to the 1896 BAAS meeting on the work that such a laboratory would do and how it should be managed.[104]

The similarity of the language in Douglas Galton's 1895 speech on the PTR suggests that it was almost certainly inspired by the correspondence from Lodge and others on the BAAS committee with Francis Galton in early 1893. Douglas Galton had little direct connection with Kew Observatory. By 1895 he had retired from a distinguished career in civil engineering. He had, though, served as a secretary of BAAS since 1870 so would have been well informed about Kew Observatory. Yet probably crucial to his 1895 speeches was his direct family connection with Kew through his cousin Francis. We have no recorded correspondence on this issue between the Galton cousins, but it is plausible that Francis suggested to his cousin that the possibility was being mooted of turning Kew into a national laboratory. That nothing further happened with regard to the NPL on the Kew Committee or the BAAS committee for more than two years after February 1893 makes it likely that Francis Galton bided his time until his cousin became BAAS president. It is also possible that Francis Galton took the initiative in 1893 to develop Kew into a national laboratory because he saw it as a way of securing the observatory's financial future. In the 1870s he had played an important role in introducing the testing of clinical thermometers when Kew needed the money. Now, if it were to be turned into a general laboratory that did a full range of standardization, including work for the burgeoning electrical industry—possibly backed up by government grants—Kew potentially had a new and much more lucrative career.

At the September 1895 meeting, BAAS followed Douglas Galton's suggestion and reconvened its committee on a national laboratory. This time it was larger—fourteen members instead of ten in 1891—and was chaired by Douglas Galton himself. It now contained three influential Kew Committee members: Foster, Rücker, and Francis Galton—a further hint of a partnership between the Galton cousins. When the committee reported back a year later, it recommended the establishment of a national laboratory supported principally by the government. Importantly, it briefly summarized the work currently being carried out at Kew and noted that "the present work of the [Kew] Observatory is therefore of a character which is strictly consistent with a large portion of the work which would find a place in a national physical laboratory."[105] The committee started a petition headed by a memorandum that expressed the need to found "a National Physical Laboratory" similar to the Berlin PTR and stated, "The undersigned give their general approval to the scheme for making Kew Observatory the nucleus of such an Institution." The memorandum ended by urging that the government find the funds for the foundation and maintenance of a national laboratory. The petition was eventually signed by sixty prominent scientists, mainly physicists, including both Galtons.[106] Again, the phraseology was identical to that in the 1893 correspondence: Kew as the "nucleus" of a national physical laboratory.

Douglas Galton made the first move toward presenting the petition to the government. The following month he sent the petition to the secretaries of the Royal Society, along with a letter saying that the BAAS General Committee had advised the association's council to seek the approval and help of the Royal Society and other learned societies in impressing upon the government the importance of enlarging Kew Observatory into a national physical laboratory.[107] Galton strengthened his case with something that gave a useful urgency to the whole project: a proposal from the town of Richmond to acquire part of the Old Deer Park, the site of Kew Observatory, for civic purposes, "which would probably put an end" to the idea of obtaining the land needed to enlarge the observatory into a physical laboratory. He now asked Royal Society secretary Michael Foster to organize very quickly a deputation from the Royal Society to the government.[108] There was an unwritten implication here: no land in the Old Deer Park meant that there could be no NPL, because building a new laboratory from scratch somewhere else would cost a great deal more, making it much harder to convince the government to fund the project. Just two days after Douglas Galton's letters to the Royal Society, his cousin informed the Kew Committee of these moves. The committee passed a resolution approving "generally" of the idea of building a national physical laboratory connected with Kew and asked to be kept informed of what was proposed to be done next in this regard. Francis Galton quickly communicated the Kew Committee's approval back to the Royal Society.[109] The Kew Committee's concurrence in the project is not surprising: of the five members present at the October 1896 meeting, four were signatories to the BAAS petition of the previous month. Moreover, the committee's members had known all about the possibility of turning Kew into a national physical laboratory ever since Francis Galton, as chairman, had informed them about it in February 1893.

Douglas Galton's request for the Royal Society's assistance was read at the next Royal Society Council meeting on 5 November. The council resolved to reply to Galton that it was "wholly in sympathy" with his proposal and agreed to form a joint committee with BAAS and other learned societies "to consider the matter, and to take such action as they may find desirable." The council nominated Joseph Lister (president of the Royal Society), William Thomson (known as Lord Kelvin since 1892), Lord Rayleigh, Arthur Rücker, and Francis Galton to serve on this joint committee.[110] The joint committee formed a twenty-eight-strong deputation that visited the prime minister, Lord Salisbury, on 16 February 1897. In addition to many leading physicists, the deputation included the chemist William Ramsay and engineers such as William Ayrton and William Preece. On 3 August, the Treasury agreed to set up a government committee with a brief "to consider and report upon the desirability of establishing a National Physical Laboratory" for standardizing scientific instruments, for establishing and preserving measurement standards, and for determining physical constants and data for science and industry. The Treasury also requested the committee "to report whether the work of such an institution, if established, could be associated with any testing or standardizing work already performed wholly or partly at the public cost"—an indirect reference to Kew Observatory as well as the Board of Trade's electrical standards laboratory.[111]

The government committee was chaired by Lord Rayleigh; Moseley has speculated that Rayleigh's class connections with the government—Arthur Balfour, the First Lord of the Treasury, was Rayleigh's brother-in-law—might have helped the proposal gain support in government circles.[112] Two committee members, Courtenay Boyle of the Board of Trade and Robert Chalmers of the Treasury, were senior government officials. The others—Andrew Noble, John Wolfe Barry, William Roberts-Austen, Arthur Rücker, Alexander Siemens, and Thomas Thorpe—all represented science, engineering, and industry. Rücker was the sole member of the Kew Committee represented. As with the Devonshire Commission and the commission of 1877 on the Meteorological Office, the committee's hearings were somewhat internal, as many of the thirty-two witnesses from whom it heard evidence in the ensuing months had been on, or at least associated with, the BAAS committee that had advocated a national laboratory in the first place: they included both Galtons, Oliver Lodge, and Richard Glazebrook. Four of the witnesses—Francis Galton, Robert Scott, George Carey Foster, and Richard Strachey—were Kew Committee members. The government committee also interviewed Charles Chree, the superintendent at Kew.[113]

The Treasury published its report on 6 July 1898. The report recommended that "a public institution" for standardizing instruments, testing materials, and determining constants "should be established by extending the Kew Observatory in the Old Deer Park, Richmond." The report specified that the existing buildings at Kew should be improved and new buildings erected "at some distance from the present Observatory."[114] It is striking how the idea—the very phrase—of establishing a national laboratory by "extending the Kew Observatory" had emerged from the government's report unchanged since the initial correspondence in the early 1890s. For example, in February 1893, Arthur Schuster had suggested that such a laboratory could be started "if the work of the Kew Observatory could be extended in certain directions." The same language appears in Douglas Galton's 1895 presidential address to BAAS in which he had called for "the extension of the Kew Observatory."

Three months later, the Treasury informed the Royal Society that the government was prepared to adopt the report's recommendations. The government offered to pay £12,000 for new buildings on the Old Deer Park site in addition to the existing Kew Observatory, plus a grant-in-aid of £4,000 per annum "for 5 years certain." Otherwise, the government stipulated that the existing Gassiot Trust income of around £470 per annum should contribute toward the new laboratory's running costs, and it followed the report's recommendation that the standardization work undertaken should be self-supporting through fees—exactly as had been the case at Kew since the 1850s. The Treasury even recommended, in view of the new large endowment of £4,000 per annum, that the £400 currently being paid each year to Kew Observatory by the Meteorological Council should now be stopped.[115]

The latter recommendation annoyed members of the Royal Society Council, who objected to it in the Royal Society's official reply to the Treasury on 28 November, pointing out that in return for the annual £400, Kew provided services for which the government would have to pay if they were carried out somewhere else. In all other respects, however, the Royal Society Council was happy to accept the government's offer. In their reply, the council members confessed that they "cannot conceal from themselves" that the money offered was less than they had hoped for but said that as the NPL was to be in the form of "an extension of the Kew Observatory" (once again, the language of "extension"), the money would "greatly increase the utility and range of the work there done." They informed the Treasury that the Royal Society had appointed a committee to discuss the details of the new laboratory with the Treasury.[116] The proponents of the NPL knew, as they had known since the early 1890s, that a modest extension of Kew Observatory was the best that they could hope for in the circumstances—indeed, it was all that the BAAS committee had applied for. The artillery expert Sir Andrew Noble likely spoke for many when he recommended accepting the government's offer, in the hope that once the NPL had proven its "utility," "a future Chancellor of the Exchequer may see his way to be a little more liberal."[117]

The committee of seven appointed by the Royal Society Council to negotiate with the Treasury included two Kew Committee members: Rücker and Adams.[118] Early in 1899, its members drew up a "scheme of organization" that detailed how the NPL was to be managed. The laboratory was to be run on a day-to-day basis by an executive committee, whose twenty-three members were to represent industry and engineering as well as physics. A similar balance of members was to make up a "general board" that would provide overall direction. Six members of the executive committee were to be from the "members of the Kew Observatory Committee at the time when the Kew Observatory is incorporated into the National Physical Laboratory." Two of these Kew Committee members were to retire from the executive committee every two years, and the vacancies thereby left would not be filled from within the former Kew Committee.[119] Thus the committee that had run Kew Observatory for half a century was absorbed into—and ultimately dissipated by—the NPL, which the Kew Committee had done so much to give birth to. There only remained the legal formality of winding up the Kew Committee itself, given that the committee had been incorporated under the Companies Act in 1893. This procedure was overseen by the Royal Society's treasurer, Alfred Bray Kempe, a Cambridge mathematician who became a barrister and put his financial acumen to the benefit of the Royal Society. The Kew Committee was wound up at two extraordinary general meetings, held in late 1899. Its property and assets were transferred to the Royal Society.[120]

No member of the Kew staff was made redundant after the NPL formally took charge of Kew on 1 January 1900: the NPL's annual report for 1901 records practically the same number of staff at Kew as there had been during the last years of the Kew Committee.[121] In any event, however, Charles Chree did not become director of the new NPL—even though he seems to have been chosen as superintendent in 1893 partly with the possibility in mind that Kew might become a larger organization. The executive committee instead chose Richard Glazebrook, deputy director of the Cavendish Laboratory and a member of the original BAAS committee set up to discuss the possibility of a national laboratory. We do not know why Glazebrook was chosen. The minutes of the executive committee record that each of the members present at its first meeting in May 1899 suggested a name as a candidate for director. By mid-June, the list thus generated had been short-listed to two names, neither of them specified.[122] The earliest official record that we have of Glazebrook being chosen is in the executive committee minutes of 5 July 1899, though as early as 7 June Courtenay Boyle wrote privately to Rücker that "I doubt if we can give Glazebrook [£]1400—at any rate I think we must use all our efforts to find some other solution first."[123] Thus just one month after the executive committee came into existence, Glazebrook was the favored candidate. That Glazebrook was chosen instead of Chree might well have been due to the influence of former Cavendish Laboratory director Lord Rayleigh, who was now vice-chairman of the executive committee. Glazebrook had been Rayleigh's preferred candidate to succeed him to the headship of the Cavendish in 1884, but to the frustration of both, J. J. Thomson was appointed.[124] Now, in 1899, Rayleigh would have wanted to see "his" man become director of the new national laboratory. Glazebrook also had much more administrative and financial experience than Chree. In addition to being assistant director of the Cavendish, in 1895 he had become bursar of Trinity College, one of Cambridge's wealthiest colleges.[125]

In November 1899 the Kew Committee, in one of its final meetings, authorized the NPL Executive Committee to convert a room on the first floor of the observatory into accommodations for Glazebrook. In February 1900, the Royal Society Council accepted the Office of Woods's offer of a fifteen-acre plot in the Old Deer Park for the use of the National Physical Laboratory.[126] All seemed set for work to begin on new buildings on the Old Deer Park site. Glazebrook did briefly operate from Kew Observatory: some of his earliest letters as director of the NPL bear the same address as Kew Observatory (figure 5.2). In the spring of 1900, however, some influential local residents had objections raised in Parliament about the disruption that the new buildings might cause.[127] By the late nineteenth century, the Old Deer Park was being increasingly used as a semirural location for leisure pursuits, notably by the Mid-Surrey Golf Club. It is easy to see why many local people would have been perturbed by the plans to build a complex of buildings here, one of which was to house a turbine to generate electricity for the new laboratory. Over the following months, these objections became serious enough for the government to reconsider using the Old Deer Park as a site. By late October, the government "felt bound to consider whether any alternative site could be secured which would be reasonably satisfactory for the purposes of the Laboratory."[128] On 24 October, the NPL Executive Committee considered the government's offer of such an alternative site: Bushy House in Teddington, Middlesex, which was conveniently

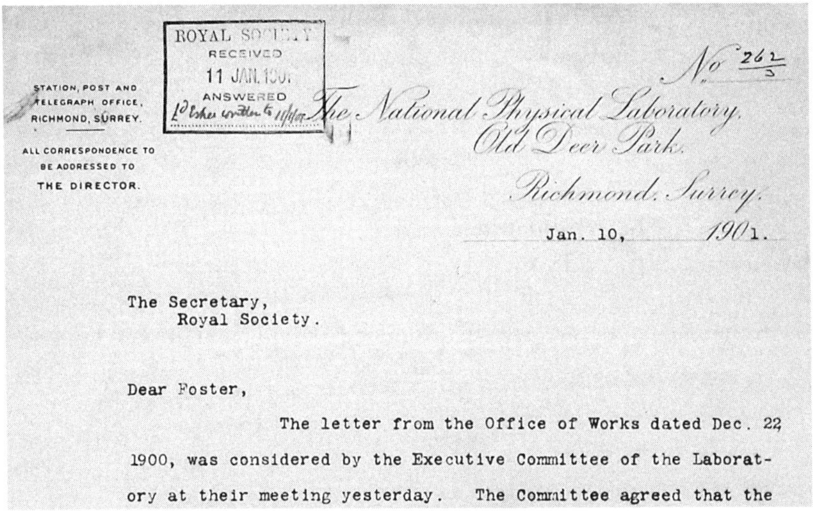
ROYAL SOCIETY
RECEIVED
11 JAN 1901
ANSWERED

STATION, POST AND TELEGRAPH OFFICE, RICHMOND, SURREY.

ALL CORRESPONDENCE TO BE ADDRESSED TO THE DIRECTOR.

No. 262

The National Physical Laboratory,
Old Deer Park,
Richmond, Surrey.

Jan. 10, 1901.

The Secretary,
Royal Society.

Dear Foster,

The letter from the Office of Works dated Dec. 22 1900, was considered by the Executive Committee of the Laboratory at their meeting yesterday. The Committee agreed that the

FIGURE 5.2. Part of a letter from Richard Glazebrook, NPL director, to Michael Foster, secretary of the Royal Society, 10 January 1901 (RS.MS.538). Note the letterhead giving the address of the National Physical Laboratory as 'Old Deer Park, Richmond, Surrey"—the address of Kew Observatory. Photograph by Lee Macdonald; used with the permission of the Royal Society.

accessible to the west of London, like Kew. Ironically, Bushy House was now in a similar situation to the Kew building in 1841: it was Crown property, most recently leased to the exiled French royal family, yet had been lying unused since 1897. The executive committee considered that "a reasonably satisfactory" NPL could be built around Bushy House and did not recommend that the Royal Society oppose the Treasury's changed offer.[129] During the course of 1901, the interior of Bushy House was converted into laboratories, and part of one floor became Glazebrook's new accommodations. The NPL finally moved to Teddington in September 1901, leaving the magnetic and meteorological work and, initially, instrument testing, behind at Kew.[130]

CONCLUSION

In his 1909 autobiography, Francis Galton devotes about one-third of a chapter to Kew Observatory and his relations with it. Yet he glosses over the role of Kew in the origins of the National Physical Laboratory with a few brief sentences that could have been derived from the published sources available at the time. Kew, according to Galton, "was wholly unequal in its scale to the rapidly growing requirements of the day. This feeling found expression in the Anniversary Address to the British Association in 1895, by my cousin Sir Douglas Galton; powerful support was given to his suggestions and efforts, and finally the Kew Committee was merged into the much larger and more important National Physical Observatory [sic], under the directorship of Mr. Glazebrook, which swallowed at a single gulp the whole of our thrifty savings."[131] Galton might have been motivated by feelings of respect toward his cousin, who had died in 1899. But crucially for historians, his summary perpetuates the myth that Kew Observatory was just a convenient first site for the NPL, or at most a Victorian precursor that did some of its work on a small scale and had little relation to the twentieth-century NPL. Indeed, it is possible that Galton's widely read memoirs helped to create the myth. More than two decades later, during a lecture to the NPL, Richard Glazebrook claimed that in the political climate of the 1890s, it "seemed impossible" to obtain government funding for a national laboratory and that the BAAS committee appointed after Lodge's 1891 speech "lapsed without taking further practical action."[132] This story of the laboratory being dismissed as a lost cause in the early 1890s, until Douglas Galton's happy intervention propelled the government into at last providing state support for a national laboratory, has largely been followed by later historians. Both Glazebrook's and Francis Galton's reminiscences, as well as more recent accounts, marginalize the role of Kew Observatory in the birth of the NPL.

It should be clear from this chapter that, far from being merely incidental to the inevitable rise of the NPL, Kew was central, in fact essential, to the NPL's origins. In the political circumstances of the 1890s, the NPL might never have become a reality without it being first presented to government as an extension of the existing Kew Observatory. It is fair to say also that the Kew Committee was instrumental in giving birth to the NPL. The Kew Committee was important not only in the behind-the-scenes moves from the early 1890s onward to establish a national laboratory but also in changing the nature of Kew Observatory itself—in many ways it came to resemble the proposed laboratory and enabled the 1896 BAAS report on the proposal to comment accurately that the work at Kew was "of a character which is strictly consistent with a large portion of the work which would find a place in a national physical laboratory."[133] This was dramatically symbolized by the instigation in the mid-1890s of research into platinum thermometers and high-temperature thermometry more generally so that by 1896, an unofficial NPL thermometry department existed several years before the inauguration of the NPL itself.

Kew did some important research in geomagnetism and meteorology in the years after 1885, but neither the Kew Committee nor the superintendent directed that research in the way that they had done under BAAS. Gone were the days when Kew led the way in Balfour Stewart's all-embracing "cosmical physics" that encompassed geomagnetism, meteorology, and sunspots in a single theory of the universe. Geomagnetic and meteorological research at Kew was now largely done in the service of other institutions. Charles Chree's own research into terrestrial magnetic variations, while ultimately a vindication of Sabine's discovery of the correlation between the sunspot cycle and terrestrial magnetism, was the work of a lone scholar and not the director of a revived Magnetic Crusade.

By the early 1890s, therefore, Kew Observatory was even more of a laboratory of service to industry and government than it had been in 1885. It was also effectively a business, as symbolized by its incorporation under the Companies Act. In its everyday running, it was largely self-supporting—just as the government expected the standardization work at the new NPL to be. Finance was still the main motive for the growing program of standardization work, already Kew's most important source of income by 1885. Not only the amount but the range of instruments tested at Kew increased enormously over the next fifteen years so that by 1900, Kew was already a central standardization laboratory whose monogram and certificates leading instrument makers were proud to display on their wares. Although some standardization work was done elsewhere—at the Cavendish Laboratory and the Board of Trade's standards department—the standardization service at Kew was the largest and most comprehensive in the British Isles. Some of this expansion of standardization at Kew was in response to external demand, but the work's potential to earn the observatory lucrative income meant that the Kew Committee seldom hesitated to meet that demand.

This increased emphasis on standardization meant that it was an easy step from Oliver Lodge's 1891 speech on the need for a national laboratory to the idea first mooted privately in 1893 of making Kew the "nucleus" of such a laboratory. The discourse of Kew as the nucleus of a larger laboratory can be traced from the correspondence of early 1893 through to Douglas Galton's 1895 speech and the proposal to approach the government a year later. The idea of the NPL as an "extension" of Kew runs with similar consistency from 1893 right through to the Treasury report of 1898 recommending the establishment of a national physical laboratory. The NPL might have remained quite literally an extension of the Kew Observatory buildings had it not been for local objections to the proposed new buildings in the Old Deer Park in 1900, by which time the NPL was already in existence at Kew.

Therefore the NPL was neither established nor conceived as a new laboratory from the bottom upward, backed by generous state support much like the PTR. Rather, it grew out of an existing institution in which the need to make money constrained it into becoming a laboratory that carried out an important part of the work that would be done at the NPL. This "extension" of Kew Observatory into the NPL was thus very much a part of the laissez-faire world of the 1890s and not that of twentieth-century state-supported science.

6

"An Epoch in the History of Kew"

THE END OF THE VICTORIAN KEW OBSERVATORY, 1900–1910

The opening of the Magnetic Observatory at Eskdalemuir and the transference of the magnetic work thither marks an epoch in the history of Kew which suggests a reconsideration of existing arrangements.

WILLIAM NAPIER SHAW TO RICHARD TETLEY GLAZEBROOK,
7 MAY 1909

IT IS TEMPTING TO THINK OF THE ESTABLISHMENT OF THE NATIONAL Physical Laboratory in 1900 as the end of an era for Kew Observatory, just as it is easy to think of the NPL suddenly coming into existence as a twentieth-century institution. But to tell the full story of Kew as a Victorian scientific institution, it is necessary to include the years up to 1910, at the end of which the management of the observatory was transferred from the NPL to the Meteorological Office. It was during these years that Kew gradually lost the essential characteristic that it had had throughout practically the entire Victorian era: that of a national institution at which several observatory sciences were practiced. By the early 1910s, the last of its Victorian aspects had disappeared and it had become recognizably a twentieth-century scientific institution, which mostly specialized in one science only.

The story ends conveniently just before the start of the First World War, which saw the final breakdown of the Victorian social and political order, along with a spectacular increase in state support for science—principally, but not exclusively, military science. This led to the beginnings of what David Edgerton has claimed to be a "warfare state," meaning the domination of the mid-twentieth-century British economy by military spending, including investment in military science.[1] But the fact that the end of the Victorian Kew Observatory coincided almost exactly with this major turning point in British and European history does not mean that its transformation should be confused with it. All the essential changes that turned it, as we can see with hindsight, into a twentieth-century institution occurred before 1914 and owed their origins and evolution to the scientific and economic conditions prevailing in Britain during the first years of the century.

This final chapter shows how this transformation of Kew Observatory happened. The first section describes how Kew kept many of its pre-1900 aspects, such as its status as a world center for instrument standardization, right up to 1910 and even slightly beyond, but argues that its overall significance began to be eroded by its being a small part of the rapidly growing NPL. Kew's significance was further undermined by an enforced removal of the magnetic instruments to a new site in Scotland, as described in the second section. By 1910, a central function of the Victorian Kew Observatory had gone. The third section describes how Kew ceased to be a multifunction observatory with the transfer of its management from the NPL to the Meteorological Office and, in return, the removal of instrument standardization to the NPL's Teddington site. With standardization and geomagnetic observation gone, Kew became a specialized institution dedicated mostly to meteorology, which it remained until its closure in 1980. The final section briefly reviews how Kew worked after 1910 and compares the twentieth-century Kew Observatory with that of the nineteenth century. I conclude that to understand the 1910 handover of Kew from the NPL to the Meteorological Office, we need to consider the whole period from 1900 to 1910. Kew gradually lost its Victorian characteristic due partly to the changing scientific culture of Edwardian Britain, notably the new meteorology that was emerging around the turn of the century. Increasing government involvement in scientific institutions was also an important influence on Kew after 1900 and also in the establishment of the "new" Kew in Scotland, the Eskdalemuir magnetic and meteorological observatory. Yet the government's continued adherence to nineteenth-century laissez-faire economic policy also remained an important part of the culture in which Kew operated in its last years as a Victorian scientific institution.

THE "OBSERVATORY DEPARTMENT" OF
THE NATIONAL PHYSICAL LABORATORY

After the National Physical Laboratory, with Richard Glazebrook as director, moved to Teddington in 1901, Kew Observatory became known as the "Observatory Department" of the NPL. It retained most of the functions it had had before 1900, but Chree—now designated superintendent of the Observatory Department[2]—saw his status and independence much diminished. He no longer reported directly to a Kew Committee composed of gentlemen-scientists, military officers, and some university physicists. The NPL had an extra layer of management: Chree now reported to Glazebrook, who himself reported to the NPL Executive Committee—a much larger body than the Kew Committee and overwhelmingly made up of professional scientists and engineers, with a small smattering of industrialists. Glazebrook was the only member of the NPL staff in attendance at executive committee meetings in the 1900–1910 period. The committee's minutes never record Chree as present, the way he was at meetings of the Kew Committee. In 1912, Chree reminisced how before 1900, he had visited Kew Committee chairman Francis Galton at his home each month to go through the business of the next committee meeting.[3] Those gentlemanly days, when Chree had enjoyed regular, one-to-one contact with the governing body's chairman, were now gone.

Indeed, Kew Observatory as a whole became subordinate to, yet at the same time isolated from, the main NPL, four miles away in Teddington, especially when the Teddington site began to expand with more departments, buildings, and people. While the number of staff at Kew remained almost static between 1900 and 1910, the workforce at Teddington grew dramatically over the same period. At the end of 1902, the Observatory Department had 18 staff, practically unchanged since the 1890s, while Teddington had two departments, physics and engineering, with a total of 28 staff. In January 1910, six months before it became part of the Meteorological Office, Kew still had just 19 staff, while Teddington had four departments and a total workforce of 103.[4] Teddington also expanded physically, with a number of large new buildings going up around Bushy House by 1910, while no new building work was undertaken at Kew. Similarly, the Observatory Department's annual reports, although similar in layout and length to the pre-1900 reports, became a progressively smaller part of the larger NPL reports, and Kew became just one matter among many discussed at executive committee meetings.

By 1910, therefore, Kew Observatory was a small outlying department of a much larger organization. A feeling of being both isolated and not in control is clear from Chree's outgoing correspondence, as in a January 1902 letter about authors' copies of papers. Chree wrote, "I understand from Dr Glazebrook when I last saw him that he was to write to, or talk with, one of the Secretaries [of the Royal Society] on the subject," suggesting that Chree did not see Glazebrook very often.[5] There is a sense that this isolation was causing Chree to feel frustrated. The tone of a 1909 letter to Glazebrook gives the impression that Chree was feeling out of touch with what was going on: "Has it been decided whether there is to be an additional telephone in my room only or in Mr Constable's as well? It was proposed to put the one in my room in during my holiday."[6] Chree had not initially been keen on having a telephone in the observatory at all. In 1899, just before control of Kew passed to the NPL, a representative of the National Telephone Company had approached Chree with an offer to connect the observatory to the telephone service over the entire London area. In Chree's view, the telephone "would be apt to prove a serious weapon of offence in the hands of London instrument makers,"[7] who were likely anxious to hear about the status of the wares they had sent for testing. But under the NPL, Chree had no option but to accept the new technology. In April 1903, the NPL Executive Committee decided that Kew was to be connected directly to the telephone exchange, with an extension being put through to Teddington.[8]

Although things changed considerably for Chree after 1900, there was a striking continuity between the Kew Committee and NPL eras in the makeup of the staff who worked under him (figure 6.1). Five of the most senior members of staff at Kew in 1899 (including Chree himself) were still there in 1910. Of these, all except Chree had worked at Kew for at least thirty years by 1910. Another senior assistant, T. Gunter, died in post in 1906, having joined the staff at Kew in 1875.[9] Two further members of staff, W. Hugo (senior assistant) and G. E. Bailey (accountant and librarian), transferred to Teddington soon after 1900. By 1910, Bailey had become the NPL's chief clerk and accountant.[10]

As before 1900, the salaries of the staff at Kew remained low compared to those of their counterparts at Greenwich Observatory. Chree enjoyed an increase in salary from £550 per annum in 1901 to £656 in 1909. By the latter date he was also receiving a pension allowance—an

FIGURE 6.1. Group photograph of Kew Observatory staff in 1906. Charles Chree, superintendent, is seated at the extreme right of the middle row. At least four of the other figures can be identified in the group photograph in figure 4.2. Photograph courtesy of the National Physical Laboratory.

improvement on his job description when he applied for the position in 1893, which had specifically stated that no pension was provided.[11] A 1904 pay review acknowledged that many of the assistants' salaries were poor, considering the jobs they held and their long periods of service. The chief assistant, Thomas Baker, was earning £325 per annum, while the salaries of the five senior assistants were capped at £200. Three years earlier, the three junior assistants were earning between £95 and £125, a figure unlikely to have changed by 1904.[12] Salaries at Kew clearly did not rise much when Kew became part of the NPL. As Moseley has described, the salary situation was no different at Teddington: due to the constant shortfalls in funding, the scientific staff of the early NPL remained poorly paid throughout the years before the First World War.[13] Under the NPL, the Observatory Department continued to employ teenage boys as clerks to do basic computing and clerical work. In February 1902, Chree sent advertisements to two local newspapers, asking for "*Boyclerks*—Age

15–17" and requesting that candidates apply in their own handwriting. The staff lists in the NPL's annual reports for the years to 1910 include around six boy clerks at Kew.[14]

In one respect, the staff of the Observatory Department was in stark contrast to that at Teddington. Whereas most of the senior assistants at Teddington had at least one university degree, none of the assistants at Kew up to 1910 had any letters after their names. Most of even the senior assistants had joined the staff at Kew as boy clerks and then slowly worked their way up through the ranks over the course of very long careers there. Gunther, for example, had started in 1875 as a boy clerk; at the time of his death in 1906 he was head of the thermometer-testing division.[15] Yet under Chree's broad supervision, these men took responsibility for inspecting the principal observatories of the Meteorological Office across Britain and Ireland and testing some of the finest scientific instruments and timepieces of the day, in addition to making and reducing the various observations at Kew. The tradition of hiring low-paid assistants with little or no formal scientific education seems to have mirrored the situation in the precision instrument-making industry in the first years of the twentieth century. Mari Williams notes that a survey conducted in 1910 by London County Council indicated that only a modest number of employees of optical firms were attending evening classes or part-time courses to further their education.[16]

A memorandum dated February 1910—very likely written by William Napier Shaw, who had succeeded Robert Scott as head of the Meteorological Office in 1900—gives some brief but vivid insights into life at Kew Observatory in the first decade of the twentieth century. The hours of work were 9:30 a.m. to 5:00 p.m., rather like those of an office. No observer was living in the building by then: only a caretaker lived there, in the basement. This was very different from the 1840s and 1850s, when John Galloway, William Radcliff Birt, and John Welsh had all lived on the premises. It highlights a problem with the Kew building that had become apparent even by the late 1880s: lack of space. Rooms that might have been made into living quarters for observers were now being used for testing the numerous types of instruments that went through Kew, with the result that "a visitor to the observatory could hardly fail to carry away the impression that the scientific staff and the caretaker are making the best of a makeshift situation."[17]

The memorandum naturally concentrated on the meteorological work at Kew that, on the surface, continued unchanged after 1900. In addition to performing routine observations as one of the Meteorological Office's principal observing stations, Kew Observatory continued to function as the office's "central observatory" in that members of its staff regularly inspected the instruments of the other principal stations to ensure that they corresponded with the standards at Kew. The inspections continued to be carried out by chief assistant Thomas Baker and one of the senior assistants, E. G. Constable, but never Chree. A 1907 memorandum, probably also by Shaw, described these inspections as "a matter of stolid routine."[18] This work of Kew as central observatory continued to be funded by an annual £400 grant from the Meteorological Office, as it had been since 1876.

After 1900, Kew retained its international status as a center for meteorological observation. In October 1908, for example, at the request of the head of the Russian meteorological service, an official from Russia's Pavlovsk Observatory paid an eighteen-day visit to Kew to compare the standard magnetic instruments and barometers in Britain and Russia. This was part of a larger international comparisons project agreed on at a recent international meteorological conference, one of a series of such conferences promoting worldwide cooperation in the field.[19] Kew remained a frequent participant in Meteorological Office projects, such as a 1901 attempt to study the formation of the notorious London fogs during winter.[20] As during the last quarter of the nineteenth century, these projects were led by the office rather than Kew. Scientific workers near the beginning of their careers frequently visited Kew to learn the techniques of observational meteorology. A notable example in 1906 was George Simpson, a young meteorology lecturer who was about to start work for the Indian Meteorological Service. Simpson served as a meteorologist on Robert Falcon Scott's ill-fated second Antarctic expedition and in 1920 succeeded Shaw as director of the Meteorological Office. In his retirement, he took over as superintendent of Kew Observatory in 1939 and remained in this position until 1947.[21]

The somewhat negative tone of the 1907 and 1910 memoranda on Kew can be explained partly by the fact that both were written to justify a proposal for the Meteorological Office to take control of Kew Observatory and for the standardization work to move to Teddington (see later in this chapter). Another factor, however, was the changing culture of the Meteorological Office in the first years of the new century. This culture change was both scientific and administrative. Shaw's predecessor at the helm of the office, Robert Henry Scott, was fundamentally a scientific administrator and a collector of data. Scott did not attempt to discover general laws governing meteorological phenomena using mathematical and physical principles. Under Scott's leadership, the Meteorological Office was rather like a public statistical service that produced data and did not have a research department that attempted broader theoretical explanations. Shaw was very different from Scott. Like Chree and Glazebrook, his intellectual outlook had been formed at Cambridge, especially the Cavendish Laboratory. He had begun working in meteorological physics while at Cambridge and was a pioneer of the new "dynamical meteorology" that used the latest principles of physics to seek general theories of how weather systems moved and worked. Soon after he succeeded Scott at the helm of the Meteorological Office in 1900, Shaw made research—including theoretical meteorology—central to the office's program of work and began recruiting university-trained physicists to senior positions, most of them Cambridge men of his own stamp.[22] In this new meteorology, observation was still important, but it had to be directed with theory in mind: mere data collection was not enough. In this respect, it resembled John Herschel's theory-driven approach, but now it had the object of accurate forecasting rather than gentlemanly scientific curiosity. Shaw's attitude toward observational meteorology is exemplified in a letter he wrote to fellow meteorologist (and Cambridge graduate) Lewis Fry Richardson, appointing him to a post: in his letter, Shaw apologized for the amount of routine observation and data reduction that Richardson would have to do.[23] Shaw would have seen Kew Observatory in this context—as an observation station whose work should be subservient to his broader research agenda rather than as a nerve center for several sciences.

The administrative side of this cultural change began in 1902, when some questions in Parliament about the Meteorological Office's performance led to a government enquiry, which reported in May 1904. In accordance with the report's recommendations, the following year the office's governing body, the Meteorological Council, was abolished and replaced with a new Meteorological Committee. This committee, confusingly, had the same name as the body that had run the Meteorological Office from 1867 until 1877. However, whereas the old Meteorological Committee and Council had mostly been made up of university physicists, gentlemen scientists, and scientific servicemen and chaired by one of their number, the new committee was to contain one representative each from the Treasury, the Board of Trade, and the Board of Agriculture, plus the hydrographer of the Admiralty and just two representatives of the Royal Society. It was to be chaired by the director of the Meteorological Office, and its members were all to be appointed by the Treasury. Each year, it was required to submit a spending proposal to the Treasury and a report to Parliament.[24] From 1905, therefore, the Meteorological Office's governing committee was effectively a government department, with as many civil servants as scientists—in contrast to the old Meteorological Council, which was appointed by the Royal Society to administer the annual government grant to the office.

These scientific and administrative changes resulted in the Meteorological Committee seeing Kew Observatory very differently from the old Meteorological Council. Although in practice the membership of the council was often different from that of the Kew Committee, there was still considerable overlap between the two bodies, and the council, being made up of Fellows of the Royal Society, was instinctively sympathetic to Kew as a Victorian multifunction observatory. Now the new Meteorological Committee had two different agendas. On the one hand, the civil servants on the committee were there to represent their departments: the Treasury's mandarins, in particular, were ever keen to limit public spending. At the same time, the scientists, headed by Shaw, believed that Kew should play its part in a larger, theory-directed observing regime. Neither party had any particular affinity for the pre-1900 conception of Kew Observatory. This would have an important bearing on the future relationship between Kew and the NPL.

The dramatic increase in the number and range of instruments tested at Kew between the 1870s and the end of the nineteenth century did not continue after 1900. The total number of instrument tests (excluding watches and chronometers) reported in 1900 was 27,569; the figure for 1910 was 35,826 (see table 6.1). The record year was 1909, which saw a total of 41,318 instruments tested at Kew.[25] Clinical thermometers remained the largest category of instruments tested, their proportion dropping only slightly from 87 percent to 80 percent of all thermometers verified at Kew. Yet the number of clinical thermometers going through Kew never rose above the peak of 25,861 in 1909; for most of the first decade of the twentieth century, the figure hovered around the 20,000 mark first achieved in 1900 and sometimes even dropped slightly below this (see table 6.1). The large jump in the testing of meteorological thermometers during 1912, the last year in which instrument tests at Kew are itemized in NPL reports, was explained by Chree as be-

TABLE 6.1: Principal Kew instrument verifications, 1900–1912

	1900	1901	1902	1903	1904
Met. thermometers	2786	3077	2733	2851	3157
Clinical thermometers	20476	20389	22912	19393	15903
Deep-sea thermometers	83	112	44	56	41
Total thermometers	23547	23786	26018	22577	19318
Percentage clinical	87	86	88	86	82
Barometers[1]	416	427	407	384	437
Hydrometers	173	120	403	353	706
Sextants	813	938	769	901	957
Telescopes	1345	2029	1678	3180	2943
Binoculars	963	669	924	1048	1027
Camera lenses	136	9	6	0	0
Milk-test apparatus	0	527	159	89	202
Watches	403	363	530	458	429
Marine chronometers	53	36	33	48	43
Total instruments	27983	29027	30987	29103	26132

Notes: [1] Including aneroids.

- Total thermometers does not include deep-sea thermometers.

Source: Data from "Report on the Observatory Department, 1900," and NPL Reports, 1901–1912.

ing due to a large number of maximum-minimum thermometers being tested for use in cotton cloth factories—the result of legislation requiring checks of humidity in such factories.[26] This was an example of external demand for instrument tests at Kew, very much in evidence during the late nineteenth century, continuing into the twentieth.

An important source of this external demand before 1900 had been the Admiralty. Demand from the military continued to increase in the new century. In fact, instruments for the military—especially binoculars, telescopes, and sextants—account for much of the rise in instrument tests at Kew after 1900. A particularly large number of telescopes are recorded as having been tested in these years, the figure reaching a peak of 5,376 in the year 1907, and the yearly numbers remained close to

1905	1906	1907	1908	1909	1910	1911	1912
3626	3875	5397	4719	4968	4999	4977	9133
16089	17518	20427	18752	25861	21844	22186	20909
56	40	70	5	50	37	3	55
19999	21844	26056	23710	31126	27527	27678	31365
80	80	78	79	83	79	80	67
387	525	467	311	403	535	398	612
530	571	480	613	728	495	678	498
1044	1096	1261	1154	1281	1376	1296	1325
3627	4657	5376	3177	5288	4053	4485	4288
554	390	787	1238	2292	1609	1727	2010
0	0	0	0	0	0	0	0
137	153	72	0	0	0	0	0
456	272	246	252	380	474	591	473
42	91	174	82	108	94	92	106
26848	29687	35018	30592	41676	36235	37002	40775

this for most years thereafter up to 1912. The peak in 1907 is likely the result of a 1905 agreement between the NPL and the War Office to test telescopes for the British army at Kew, in addition to the already large number of telescopes being tested for the Admiralty.[27] The increase in all three types of instruments being sent for testing reflects the surge in rearmament, especially naval rearmament, that began in the mid-1900s.

Just two new types of instruments were tested at Kew in significant numbers after 1900. In 1901, the NPL Executive Committee agreed that the Observatory Department should take on the testing of Stephen Moulton Babcock's milk-testing device, which was designed to detect dishonest farmers falsifying the fat content of their milk. Over 500 of these devices were tested in that year alone.[28] Eight years later, the executive committee agreed to a request from the Metropolitan Police to test stop-watches designed to deal with a problem that certainly did not exist when standardization at Kew began in the early 1850s: speeding motorists. Kew Observatory's watch- and chronometer-rating department began testing these stopwatches in early 1911, but the work had been going on for less than two years before it, together with all the other watch- and chronometer-rating work, moved to Teddington in November 1912.[29]

More standardization work moved from Kew to Teddington than was taken on at Kew after 1900. Notably, the testing of platinum resistance thermometers, carried out at Kew under John Allen Harker since the mid-1890s, moved to Teddington in 1902—making it even easier for historians to gloss over the essential role of Kew Observatory in the origins of the NPL's high-temperature thermometry department. The same year also saw the removal to Teddington of photographic lens testing, originally instigated by the Kew Committee in 1891.[30] Except for the milk-testing apparatus and police stopwatches, when the NPL began testing new types of items, the work was established at Teddington, not Kew. In 1903, the two divisions then existing at Teddington, physics and engineering, tested thirty-nine types of items, of which just three can be recognized as previously having been tested at Kew.[31] As the Kew building was already cramped, work on a larger scale would have required new buildings, which were not possible on the Old Deer Park site because of the planning objections that had forced the executive committee to look for a new site for the NPL in 1900. An October 1909 report of a joint committee of the Royal Society and NPL acknowledged this lack of space (and Kew's isolation) in making the case for turning Kew into a purely Meteorological Office establishment and moving the testing to Teddington: " . . . owing to the isolation [from Teddington] and limited accommodation at Kew, it is not possible to develop many of the tests in the manner required by modern progress."[32]

Instrument testing at Kew continued to enjoy a high international renown, symbolized by two visits to the observatory in 1901 and 1906 by representatives of the Japanese navy to learn about the testing processes.[33] The Kew Observatory monogram and certificates were so coveted that makers began to use them fraudulently. Back in 1899 the Kew Committee, in one of its final meetings, heard a report that one watch-maker had been claiming that his wares had received 144 Kew certificates, a claim "much in excess of the facts."[34] Then, in December 1904, Glazebrook received a letter from Jacob Pillischer, a well-known scientific instrument maker and the inventor of a type of clinical thermometer widely distributed in South America. Pillischer had discovered that some of the thermometers bearing his name had been forged. He had traced the offender back to Germany and filed a successful lawsuit against the firm in question. In the process, he had found that some of the fake thermometers bore the "KO" monogram, which, according to Pillischer, was "so much valued by the Medical Profession and Public." Pillischer asked Glazebrook to use his influence to protect him and other thermometer manufacturers. The NPL Executive Committee brought the issue to the attention of the Board of Trade, via the Royal Society Council.[35] To protect the Kew monogram (and also the NPL's separate monogram inscribed on instruments tested at Teddington), the executive committee resolved to register them as trademarks; the Board of Trade authorized their registration in March 1908. The Kew Committee had unsuccessfully tried to register the "KO" monogram as a trademark in 1880 (see chapter 4). The NPL's success in 1908 was likely because the application was made in the name of the Royal Society, which had more legal weight than the NPL Executive Committee or the Kew Committee.[36]

FROM KEW TO ESKDALEMUIR: MAGNETISM MOVES NORTH

As with meteorology and standardization, Kew Observatory kept its reputation as a world center for geomagnetic observation, verifying magnetic instruments for colonial and foreign observatories. It was also during these years that Charles Chree cemented his reputation as an expert in geomagnetism and its relation with solar activity. However, after 1900, it was the magnetic department that would undergo the most dramatic change—a change that would play a fundamental part in Kew ceasing to be the multifunctional scientific institution that it had been since the 1840s.

In its 1860 description of Kew Observatory, the Kew Committee had noted that "the repose produced by its complete isolation is eminently favourable to scientific research."[37] But in the late 1890s, London's expanding suburbs brought with them a problem that directly threatened the cherished magnetic work at Kew. Early in 1898, Parliament sanctioned the building of a new electric tramway between Kew Bridge and Hounslow that would pass within 1300 yards of Kew Observatory. In the belief that the tramway as proposed "would ruin the magnetic work at Kew Observatory," the Kew Committee appointed a subcommittee, chaired by Arthur Rücker, to look at ways of minimizing the effect of

the tramway on the magnetic instruments.[38] The Kew Committee managed to persuade the Board of Trade, via the Royal Society Council, to
stipulate that the tramway's power lines be thoroughly insulated and that
this be a requirement for any future tram lines that might affect Kew
Observatory. Yet in its report for 1898, the committee admitted that it
was "impossible to contemplate the future without some misgivings."[39]

These misgivings soon proved to be well-founded. By late 1901, the
tram service from Kew Bridge to Hounslow was running and was severely
distorting the traces from some of the magnetometers, so that "for the
investigation of small natural disturbances . . . the traces are absolutely
useless."[40] By then, another tramway was being built that would pass just
half a mile from the observatory. Not only would this cause even greater
magnetic disturbances, but its effect on the self-recording instruments
in the main observatory would be different from that on the instruments
some yards from the building that measured the absolute magnetic field,
making it very difficult to compare the two sets of instruments.[41] The
effect of the trams on the instruments was such that from 1902 onward,
the tables of diurnal inequalities in vertical force and magnetic incli-
nation were not included in the Observatory Department's annual reports.[42] For the first time since the early days of the observatory under
BAAS, Kew Observatory had had to curtail its magnetic reports.

Yet in the meantime, the Board of Trade had agreed to take further
action. In October 1899, it proposed to review all tramways in the London area that might have an effect on Kew Observatory and other sus-
ceptible institutions.[43] This might explain why, in the spring of 1901, the
government entered into negotiations with the London United Tramway
Company, which had built the line from Kew Bridge to Hounslow. The
government offered to give the company permission to open the line
on 4 April that year (in time for the Easter holiday, when traffic was
expected to be heavy) in return for the company paying the government
£10,000 in compensation toward the cost of removing the magnetic
instruments from Kew and reerecting them at a new site remote from
magnetic interference. By 16 April, the tramway company had formally
agreed to pay this sum in compensation.[44]

On hearing this news, the NPL Executive Committee immediately
appointed a "Magnetic Observatory Committee" (known from October
1902 as the "Magnetic Sites Committee") to consider a site for a new
magnetic observatory. This had five members: Robert Bellamy Clifton,
Ettrick Creak, John Perry, Arthur Rücker, and Arthur Schuster.[45] Of

these, all but Schuster had served on the Kew Committee before 1900 and so were familiar with the observatory's history and the environment needed for the magnetic instruments. The challenge before the committee was to select a site that not only avoided tramways but was also well away from any railways, some of which were starting to be electrified at the turn of the twentieth century. For this reason, the committee quickly rejected several possible sites in southern and southwestern England in favor of one in Exmoor (Devon) and two areas in southern Scotland: one between Dumfries and Ayr, the other near the English border, to the north of Carlisle. Exmoor was quickly eliminated due to the future possibility of railways and iron ore mining in the area. Of the two Scottish sites, the committee considered the border area the most free from magnetic disturbance. Here, a good road ran through the area between the towns of Langholm and Selkirk. The report then noted that "there is a village on the road in question—Eskdale-Muir—about 12 miles from Langholm, and from Eskdale-Muir onwards, for some 20 miles, the road is at least 9 miles from any railway." Eskdalemuir also had the advantage of being easily reached by bicycle from Langholm station. Even at this stage, the committee expressed its preference for Eskdalemuir.[46]

In January 1903, Glazebrook, Sir F. Hopwood of the Board of Trade, and Thomas Heath of the Treasury agreed to enter negotiations with a local landowner, the Duke of Buccleuch, about acquiring a plot of land at Eskdalemuir. Four months later, Chree visited the site with his chief assistant, Thomas Baker, and confirmed it to be free from any significant local magnetic disturbances. By the end of July, the government's Office of Works had drawn up plans for the new observatory. The site's remoteness meant that the observatory's employees and any visitors would have to be accommodated on-site: living nearby, as the Kew staff did in Richmond, was not an option. The superintendent was to have a separate house, with a servant's bedroom and two bedrooms for visitors. Negotiations with the Office of Works fixed the cost of the new buildings at £15,000. In December the Magnetic Sites Committee, the NPL Executive Committee, and the Royal Society Council all agreed on the plans.[47] The Royal Society's letter to the Treasury emphatically asked the government to proceed with building the new observatory, "as it is most desirable that it should come into operation with the least possible delay in order to avoid a break in the records." In January 1904, the Treasury confirmed that provision for a new observatory at Eskdalemuir would be made in the financial estimates for 1904–1905.[48]

Establishing the new magnetic observatory proved to be a slow pro-cess—due partly to Chree discovering, in November 1906, that the stone used for the new buildings contained iron in sufficient quantities to ren-der the buildings unsuitable for magnetic observations. The Office of Works had gone ahead with its own choice of stone without consulting the Magnetic Sites Committee, despite that committee's instructions to do so. The office agreed to replace the stone with a nonmagnetic type of stone.[49] In the summer of 1907, Chree made further test observations and finally approved the proposed magnetograph rooms as suitable. By the end of that year, the NPL was able to report that the new observa-tory was "practically complete" and that it had appointed a staff to run it, consisting of a superintendent and two assistants.[50] The observers duly arrived in May 1908 and almost immediately began making obser-vations. Yet the problems did not end there. The wet climate of south-west Scotland, combined with the exposed location of the Eskdalemuir site, quickly led to the basement magnetic rooms being affected by damp, compelling the observers to temporarily remove the magnetographs to a room originally intended as a general laboratory. The continued delays caused the executive committee considerable anxiety as magnetic condi-tions at Kew continued to worsen. Not until May 1910 could the NPL report that the Eskdalemuir buildings were in a suitable state for mag-netic observations.[51]

In November 1906, with the new observatory under construction, the NPL Executive Committee decided that Chree should have overall direction of the magnetic observations at both Kew and Eskdalemuir and also that the superintendent of the new observatory, when appointed, "should be a man of scientific standing and initiative."[52] George Walker (1874–1921), who started work as superintendent at Eskdalemuir in May 1908, could have been a younger version of Chree. Raised in Aberdeen, he was educated at that city's Robert Gordon College, the Royal Col-lege of Science in South Kensington (where he came to the attention of Arthur Rücker), and Trinity College, Cambridge. He graduated Fourth Wrangler in mathematics in 1897 before winning Cambridge Universi-ty's coveted Smith's Prize and then doing further postgraduate study at Göttingen. In 1903, he became a lecturer in physics at Glasgow Univer-sity, a position he retained until his appointment to Eskdalemuir. Like Chree, his research interests before his observatory appointment were primarily in theoretical physics rather than observational work in geo-magnetism and meteorology.[53]

It was clear from the start of his appointment that Walker would be doing more than just magnetic observations. In December 1902, Arthur Schuster had suggested that, because geomagnetism and meteorology were very closely related, the Meteorological Council (as it was still known before 1905) should be involved in the planning of the new observatory. As a consequence, a committee of the Royal Society had recommended that in addition to the magnetic observations, the new observatory should carry out experimental work in atmospheric electricity and other areas of meteorology.[54] From June 1908, Walker and his assistants began a regular program of meteorological observations at Eskdalemuir and reported the results to the Meteorological Office in the same manner as Kew and the office's other principal observatories. In September, they commenced seismological observations using a Milne seismograph.[55]

Right from 1908, therefore, Eskdalemuir was effectively a new Kew Observatory, minus the standardization work, but otherwise a multifunctional institution, working in geomagnetism, meteorology, and seismology. Furthermore, this "new" Kew was directed by a physicist with an identical scientific background to that of Charles Chree. Even the type of superintendent the NPL was looking for—"a man of scientific standing and initiative"—was a direct echo of the 1893 description of the vacant position at Kew that was filled by Chree: someone with "scientific status." Before heading for Eskdalemuir, Walker spent several months at Kew in early 1908, learning the observational regime there.[56] That Eskdalemuir was a replacement for Kew undermined Kew's raison d'être as a multifunctional observatory—it had already lost considerable status since the turn of the century when it had become part of the National Physical Laboratory and the tramways had compromised the accuracy of its magnetic instruments.

STANDARDIZATION MOVES TO TEDDINGTON, 1907–1912

In November 1907, as Eskdalemuir Observatory neared completion, Shaw presented to the Meteorological Committee a proposed program of work for the Meteorological Office in the 1908–1909 period. Prominent among the proposals was a "reconsideration" of the relationship between the Meteorological Office and the National Physical Laboratory with regard to Kew Observatory. As a department of the NPL, Kew was governed by the laboratory's executive committee, yet Shaw pointed out that, in contrast to the days of the Kew Committee, the NPL Executive

Committee contained no representative of meteorology or geomagnetism. Of the £400 that the office paid to Kew Observatory each year, £250 was for it to operate as one of its principal observatories, while the remaining £150 paid for Kew to act as the office's "central observatory," with responsibility for inspecting the other principal observatories. The meteorological observatory inspections, which had formerly been carried out by Kew's superintendent (George Whipple), had now become "a matter of stolid routine" and were being done by an assistant at Kew. Shaw thought that the work now being done by Kew as the central observatory represented very poor value for the £150 the office was paying each year. More importantly, there was an implication. The lack of meteorological representation on the executive committee, in addition to the complaint that Kew was giving a poor return on the Meteorological Office's investment, meant that the office now had little control over its own allegedly central observatory. This meant a great deal to Shaw, who, as already noted, had much more ambitious scientific goals than his predecessor. Shaw claimed to have no definite plan as to how to change this state of affairs but recommended that the Meteorological Committee should give it consideration over the coming year.[57]

In any event, it was not long before Shaw made further moves in this direction. In January 1908, Glazebrook wrote to Shaw, asking for permission to use at Eskdalemuir some Meteorological Office apparatus currently stored at Kew. Glazebrook's letter began, "As I think you are aware, the suggestion has been made (and you were good enough to give it favourable consideration) that some at any rate of the meteorological apparatus belonging to the Meteorological Office . . . should be erected at Eskdalemuir. . . ."[58] Clearly, there had been some informal contact between Glazebrook and Shaw about Kew and Eskdalemuir before Glazebrook wrote his letter. Shaw seems to have used Glazebrook's letter as a pretext for again raising the issue of the office's relationship with Kew Observatory. A few days later, he presented the Meteorological Committee with another memorandum, which started as a response to Glazebrook's request and then went on to raise wider questions. Shaw again pointed out the problem of control, saying that while in other countries the central meteorological observatories were usually directly governed by those countries' national meteorological services, the British Meteorological Office did not have this control over Kew. The connection between Kew Observatory and the office had "gradually weakened" to the point that Kew could not now "properly be regarded as a central observatory for the office system." Having Meteorological Office instruments tested at Kew was not only expensive—it also involved time-consuming negotiations with the NPL. The impending opening of Eskdalemuir Observatory now brought an opportunity to ask whether Kew could be brought into such a relationship with the office that it could once more serve effectively as a central observatory. Shaw suggested placing the staff of Kew Observatory under the control of a joint committee of the Meteorological Office, the NPL, and the Royal Society (effectively a revived Kew Committee, though Shaw did not put it that way) but claimed that Glazebrook did not want this to happen "until the chief part of the testing work now done at Kew is transferred to Bushy." Such a transfer would require government funding, and Glazebrook, claimed Shaw, did not want to apply for such money just yet.[59]

Shaw's memorandum shows that from almost the beginning of the negotiations, Glazebrook was thinking of moving the instrument-testing work from Kew to Teddington. That Glazebrook did not wish to apply for funding immediately is likely due to the NPL's chronic financial shortfalls and years of difficult negotiations with the Treasury since 1900. In 1904, for example, the Treasury had reluctantly agreed to give a grant of £28,750 for new buildings at Teddington only after Lord Rayleigh's family connections had enabled him to discuss the proposal in person with the prime minister, Arthur Balfour.[60] In further correspondence, Glazebrook agreed that the Meteorological Office should have control of its central observatory at Kew and confirmed that he was willing to co-operate with Shaw's proposal, provided that the requisite funds could be found to move the instrument testing to Teddington. Glazebrook clearly agreed in principle with Shaw's proposals. As Shaw remarked in his reply to Glazebrook, "I think we are really agreed upon these matters, and I hope that we shall shortly find some means of expressing our opinion in an effective manner."[61]

Neither side expressed their opinion for more than a year thereafter, possibly because Shaw was preoccupied with arrangements for the Meteorological Office's upcoming move from Victoria Street to new, purpose-built premises in South Kensington. Then, on 7 May 1909, Shaw wrote to Glazebrook, making the same proposals as a year earlier, only now giving them urgency with a proposal to terminate the annual grant of £400 that supported Kew Observatory. Shaw gave many of the same reasons as he had the previous year for reconsidering the relationship between the office and Kew. These included the Meteorological Committee's belief that the principal observatories, including Kew and Eskdalemuir, should be run directly by the office. In addition, according to Shaw, the opening of the new magnetic and meteorological establishment at Eskdalemuir "marks an epoch in the history of Kew which suggests a reconsideration of existing arrangements."[62]

Shaw had served on the NPL Executive Committee since 1903 and so was naturally in attendance at that committee's next meeting on 21 May. After his letter had been read aloud, Shaw made a statement explaining further the Meteorological Committee's position with regard to its central observatory. The committee agreed to refer his letter to the Royal Society Council and to appoint a five-person subcommittee with power to discuss the matter further with the Royal Society. In addition to Glazebrook and Shaw, the subcommittee included Rayleigh (still chairman of the executive committee), Alfred Bray Kempe (the Royal Society's treasurer who had overseen the winding up of the Kew Committee in 1899), and another former Kew Committee member, Arthur Rücker.[63] On receipt of Shaw's proposal, the Royal Society Council appointed a committee of its own to begin discussions with the NPL. This resulted in the formation of a special NPL-Royal Society joint committee, comprising Glazebrook, Shaw, Rücker, Joseph Larmor (a secretary of the Royal Society), and Archibald Geikie (president of the Royal Society), which met on 8 July 1909.[64]

As a consequence of this meeting, Shaw and Glazebrook were asked to draft a "working agreement" between the Meteorological Office and the Royal Society. The draft agreement specified that both the staff and the running of the Kew and Eskdalemuir observatories (and a smaller observatory at Valencia in Ireland) were to be controlled by the director of the Meteorological Office. The agreement also incorporated the proposal to move the standardization work from Kew to Teddington, which would free up space at Kew for a purely meteorological observatory. The research side of the work at Kew and the other observatories—that is, geomagnetism, seismology, and meteorological work other than the observations they carried out for the office—were to come under the overall direction of the Royal Society through its "Gassiot Committee."[65] This had originally been set up to circumvent a legal difficulty that had emerged in the late 1890s when negotiations were in progress to establish the NPL. When John Peter Gassiot had donated the Gassiot Trust money to the Royal Society in 1871, the trust deed had specified that its dividends were to be used to run Kew Observatory under a committee of the Royal Society. Yet the NPL's Executive Committee was to comprise representatives from government, industry, and engineering, not all of whom would be FRS. The problem was resolved by Kempe, who suggested that the Royal Society might simply appoint a special committee of Fellows that could meet once a year and confirm that the trust money was being well spent by the managers of the NPL.[66] Every year since 1900, this Gassiot Committee had simply rubber-stamped its approval of the NPL's use of the Gassiot money but had otherwise played no active role in running Kew Observatory. Under the new working agreement, the committee was to meet at least twice a year and have overall direction of the research agendas at Kew and Eskdalemuir.[67]

In October 1909, the NPL-Royal Society joint committee approved a longer report that incorporated the working agreement drafted in July. Here, a section on the requirements of the NPL suggested two alternatives. One was to give up the meteorological work and use the Kew building just for testing instruments. The other was to transfer the standardization work to Teddington and hand over the meteorological observations to the Meteorological Office. The report did not favor the first option because the NPL did not want a permanent "isolation" of part of its operations at Kew. The second option, by removing the very profitable instrument testing from Kew, would leave a shortfall in Kew's finances. New buildings would also be needed at Teddington to accommodate the testing work. On the other hand, the second option would have the advantage of bringing all the work of the NPL to one site. It is in this context that the report noted the cramped nature of the accommodations for instrument testing at Kew. The testing work could be modernized and expanded if it were given new premises at Teddington. The report recommended that the Royal Society and the Meteorological Committee jointly approach the Treasury for the extra funds required.[68]

At its meeting the same month, the NPL Executive Committee expressed no hesitation in approving the joint committee's report, including its recommendation to apply to the Treasury for funds. The NPL lost no time in preparing for moving the standardization work from Kew to Teddington. In the report of the joint committee, the NPL had conceived the transfer "as part of a general scheme for the Laboratory." Now, at its October meeting, the executive committee authorized the drawing up of plans for new buildings that would accommodate not only the work currently being done at Kew but also its expanding metallurgy division.[69]

The Royal Society and the Meteorological Committee then both ap-

proved the joint committee's proposals in principle. Some details remained to be settled—notably, how the staff at Kew would be divided
between the Meteorological Office and the NPL, given that many of the
Kew staff spent much of their time on standardization. The formula
eventually worked out between the two organizations was that in return
for keeping the lucrative testing work, the NPL would continue to pay
the salaries of the older (and therefore higher-paid) members of the
observatory staff and also provide for their retirement. On 9 December,
the Royal Society Council approved a suitably amended version of the
joint committee's proposal and an application for funding to be sent to
the Treasury.[70]

With the three organizations involved—the Meteorological Office, the
NPL, and the Royal Society—all agreed on the proposal, the only real
obstacle facing it at the end of 1909 was funding. In the joint committee's application, the Meteorological Office was asking for an extra £550
per annum to meet the running costs of the observatory after the testing work had left in addition to a one-off payment of £1,000 for some
structural alterations needed to make the Kew building more suitable
for its new role as a purely meteorological observatory.[71] The NPL, for
its part, was applying for £4,500 to accommodate the testing work at
Teddington. In its reply on 10 January 1910, the Treasury flatly refused
to pay any of the money requested by the NPL. Sir George Murray, the
Treasury's permanent secretary, bluntly informed the Royal Society that
the instrument testing must remain at Kew "for the present at all events"
and that if Chree "has hitherto supervised the testing work, My Lords are
not aware of any reason why he should not continue to give his assistance
in that work." Murray suggested that if the NPL were to pay the Meteorological Office a portion of the running costs of Kew and the salaries
of all the staff doing testing, the Treasury would agree to responsibility
for Kew being transferred to the office and also to paying the office £550
a year to run the observatory and the one-off payment of £1,000 for
structural alterations.[72]

The minutes and correspondence do not suggest that the Treasury's
response caused any particular consternation on the part of Glazebrook,
Shaw, and the other members of the governing committees of the NPL
and Meteorological Office. Their lack of concern might be explained
partly by the fact that the government had agreed to Shaw's main aim—for
the office to have control of Kew Observatory. Yet it is also important that
the application for £4,500 to transfer the standardization was incorpo-

rated within a much larger proposal for new buildings at Teddington. By the end of 1909, the NPL's scheme for new buildings had expanded to include a new administration building in addition to accommodations for metallurgy and the standardization work to be transferred from Kew. The NPL wanted a total of £30,000: this sum included the proposed £4,500 for the Kew work.[73] There is a hint that Glazebrook, Shaw, and the Royal Society were all hoping that this much bigger grant, if they ever obtained it, might provide something for transferring the testing work. This is strongly suggested by the Royal Society Council's decision on 20 January to forward the NPL's larger proposal to the Treasury. The letter accompanying the proposal commented that the Treasury's 10 January letter about Kew was under consideration and that in view of the Treasury's difficulty with regard to the capital expenditure required for transferring the standardization work to Teddington, the council wished to defer expressing its views on the NPL's larger proposal.[74]

In the meantime, Shaw, Glazebrook, and Chree met at Kew Observatory on 14 January to discuss how to put into practice the deal offered by the Treasury. One week later, the NPL Executive Committee authorized Glazebrook and Shaw to draw up a statement as to the observatory's future organization. The Meteorological Office would have overall responsibility for managing and maintaining Kew. However, as the standardization work was to remain at Kew, the salaries of eleven of the observatory's fifteen scientific staff were to be charged to the NPL, with the other four to be paid by the office—a reflection of the high proportion of standardization work being done at Kew by 1909. Even Chree's salary, though paid by the office, would contain a £200 contribution each year from the NPL.[75]

The Meteorological Committee and the Royal Society Council both approved this plan at subsequent meetings, though both organizations expressed the same reservations about dividing the running of Kew. The Royal Society informed the Treasury that in its view—and that of the NPL and the Meteorological Committee—splitting the control of Kew was "not feasible as a permanent arrangement" but that Glazebrook and Shaw were happy to try it for the time being and so the Royal Society was prepared to go along with it "as a temporary arrangement."[76] Shaw advised the Treasury that in the Meteorological Committee's opinion, splitting the staff at Kew between the office and the NPL would lead to Chree becoming the servant of two masters. Also, the meteorological work would inevitably take second place to standardization, given that the latter would be employing eleven of the fifteen scientific staff. But the Meteorological Committee considered, said Shaw, that this temporary arrangement might be used "to good account" while future plans were considered.[77] These responses to the Treasury by the Royal Society and the Meteorological Committee, sent just four days apart, suggest that a double-pronged, coordinated attack on the Treasury was being mounted by the two organizations, led by Glazebrook and Shaw. The implication of both letters accepting a divided control of Kew as a temporary measure was to put pressure on the Treasury to look at the NPL's larger application for new buildings at Teddington with kind eyes.

In the same letters replying to the Treasury, both the Royal Society and the Meteorological Committee proposed that the transfer of control of Kew Observatory from the NPL to the Meteorological Office should take effect on 1 July 1910. Shaw explained that this would allow Kew to come under the office's management in time for the start of the new decade of meteorological publications in 1911; it would also help its readiness for a meeting of the International Meteorological Committee in September 1910, when an international program for meteorological observation over the next few years was to be approved.[78] The Treasury quickly agreed to the Royal Society and Meteorological Committee's proposals, including the July 1910 transfer date.[79] The NPL honored its end of the agreement by authorizing the transfer to the Meteorological Office of four scientific staff—two junior assistants and two boy clerks—plus the observatory's caretaker.[80]

The Royal Society Council duly reconstituted the Gassiot Committee, which would supervise the research side of the meteorological work at Kew. In addition to Glazebrook and Shaw, this included Chree, George Walker, and several of the leading physicists of the time, including George Darwin, John Poynting, Arthur Schuster, and C. T. R. Wilson. As with most other Royal Society committees, its membership also included the five officers of the Royal Society: the president, the treasurer, the two secretaries, and the foreign secretary. It was a substantial committee, initially totaling sixteen members including the officers. The reconstituted committee met for the first time on 13 April 1910.[81] By 17 June, it had approved a press release describing the impending transfer of the Kew and Eskdalemuir observatories. This unequivocally announced that "the Kew Observatory will be the Central Observatory for the Office" and that all communications regarding the observatory's work for the office should be sent to the director of the office. Even more tellingly, the press release made no secret of the proposal to move the standardization work to Teddington, even though no financial provision had yet been made for this: "The work of testing instruments now carried on at Kew Observatory by the National Physical Laboratory will be removed to Teddington as soon as the necessary provision for its transference can be made."[82]

The notice appeared in the July 1910 issue of *Symons's Meteorological Magazine*.[83] From that point on, the proposal to move the standardization to Teddington was public knowledge. This was one way in which Glazebrook and the NPL Executive Committee were pressing the Treasury hard to find funding for the new buildings. Further pressure is evident in a letter from Glazebrook that same month, just three weeks after the transfer of responsibility for Kew had taken effect, in which Glazebrook asked Royal Society secretary Joseph Larmor to impress the Treasury strongly of the need for money to transfer the testing work from Kew to Teddington because "the present arrangement cannot be considered satisfactory and must be regarded as very temporary."[84]

Glazebrook saw a new opportunity to push the project forward in December 1910. He reported to the NPL Executive Committee that in order to refashion the interior of the Kew building to make it more suitable for its new role, the Office of Works was proposing to gut its entire basement. The work would also put the sextant room and barometer room on the ground floor out of action. As a result, it would be "extremely difficult" to carry on with much of the instrument testing while the building work was in progress. Glazebrook now suggested approaching the Treasury for funds to remove the standardization work to Teddington immediately, before the refurbishment was due to begin. The executive committee agreed.[85]

Shortly afterward, a private philanthropist donated a significant part of the money that the NPL Executive Committee was seeking for the new buildings at Teddington. In his July 1910 letter to Larmor, Glazebrook had reported a suggestion by Richard Burdon Haldane, the secretary of state for war (who instigated the development of military aeronautics research at the NPL), that private sources might provide half of the £30,000 being applied for by the NPL.[86] Now, in 1911, the industrialist Sir Julius Wernher agreed to provide £10,000 for the metallurgy part of the application. This contribution from private funds to a larger application—an example of what in the early twenty-first century might be termed "match funding"—seems to have persuaded the government to provide much of the rest of the money, including funds to accommodate the standardization work that was still being done at Kew. In June 1911, the Treasury agreed to provide £15,000 for new buildings at Teddington, the money to be paid in three installments over the 1911–1914 period. The new buildings were finally opened in July 1913 by the former prime minister, Arthur Balfour.[87]

The transfer of the instrument testing finally began in November 1912. That month, the watch and chronometer rating, as well as the testing of barometers and hydrometers, moved from Kew to the NPL's metrology division at Teddington. Due to delays in completing the new buildings, some of the work had to be housed in temporary accommodations: watches and chronometers were initially tested in two basement rooms in Bushy House. At the end of 1912, many instruments were still being tested at Kew, notably thermometers, sextants, telescopes, and binoculars. This work had to wait until 1913 to be transferred. Only on 19 June 1914 could the NPL Executive Committee report that the Kew standardization work had all been moved over to Teddington.[88] Just nine days later, in Sarajevo, Serb nationalist Gavrilo Princip assassinated the Austrian Archduke Franz Ferdinand and his wife, setting in motion a chain of events that would lead to the outbreak of the First World War in August 1914. That war would bring to an end the Victorian pattern of science-government relations that had operated throughout the entire history of Kew Observatory from 1840 to 1910.

It is clear that the transfer of Kew Observatory from the NPL to the Meteorological Office, and the associated move of the Kew standardization work to Teddington, were essentially the result of a gentlemen's agreement between Glazebrook and Shaw that could have been anticipated by Meteorological Committee members in early 1908. Its completion was only delayed by the Treasury's apparent reluctance to pay for the new buildings required at Teddington. Yet there is a sense from the correspondence that even the Treasury was bowing to the inevitable. In his January 1910 refusal to provide money for the buildings at Teddington, Sir George Murray qualified his opinion that the instrument testing must remain at Kew with the words "for the present at all events." Six months later, the Gassiot Committee felt confident enough to declare in public that the testing work "will be removed to Teddington as soon as the necessary provision for its transference can be made."[89]

Glazebrook and Shaw both had motives for pushing the project through. The theoretically minded Shaw wanted more control over the

Meteorological Office's central observatory: he did not want to have to go through the NPL Executive Committee, few of whose members had anything to do with meteorology. Glazebrook, for his part, saw Kew as an isolated, outlying department of the NPL. He also wanted more modern facilities at Teddington to expand the testing operations, which had reached the limit of the space available at Kew. Most importantly in this act of collusion, Glazebrook and Shaw had been colleagues for more than thirty years by 1910. Both had had long careers at the Cavendish Laboratory prior to 1900 and they were the joint authors of *Practical Physics*, a popular textbook for students in laboratory physics first published in 1884.[90] In addition, Shaw had been on the NPL Executive Committee since 1903. Therefore it is likely that well before Shaw first brought the matter before the Meteorological Committee in November 1907, he had had unrecorded discussions with Glazebrook about Kew Observatory and its relationship with the Meteorological Office. There is clear evidence for some such informal contact prior to Shaw's memorandum in February 1908. Indeed, given that both had started their respective roles at the office and the NPL at the turn of the century, it is possible that they conceived this plan even earlier and subsequently took the cue provided by the 1905 reorganization of the office or the proposal to open the new observatory at Eskdalemuir (or both) to pursue their aims.

KEW OBSERVATORY IN THE TWENTIETH CENTURY

After standardization moved to Teddington, the legacy of Kew lived on in the terminology used to describe some of the testing work. Thermometer testing was given two rooms on the ground floor of Bushy House. The room used for testing meteorological thermometers was known as the "West 'Kew' Room," while that for testing clinical thermometers was called the "South 'Kew' Room." At Bushy House, there was also a separate "marking room" for engraving thermometers that had passed the relevant tests. These thermometer-testing facilities were on a larger scale than anything at Kew. In particular, larger and more elaborate testing baths replaced the devices invented by Francis Galton in the 1870s, still known informally as the "Galtons" in 1912.[91] Test certificates for barometers, although headed "The National Physical Laboratory," bore the subheading "Kew Certificate of Examination."[92] The longest legacy was in the name of the tests for watches: the "Kew Class A"' test was only superseded at Teddington in 1951.[93]

Yet despite the assurance in the Gassiot Committee's 1910 press release that the NPL "will retain the well-known K.O. mark for use with those classes of instruments which have hitherto been tested at the Observatory,"[94] the Kew Observatory monogram, originally instituted in 1878, does not seem to have been used on all instruments tested after the move to Teddington. In 1914, the Meteorological Office authorized the NPL to continue using the Kew monogram on clinical thermometers. But William Reid cites a pair of binoculars, initially stamped with the Kew monogram and the figure "10" (meaning 1910), and then stamped in 1913, 1915, and 1917 with the NPL monogram. Even the test certificates issued by the NPL did not always acknowledge their Kew heritage. A May 1914 certificate for a sextant is headed "The National Physical Laboratory," with the NPL's monogram at the top, and contains no reference to Kew Observatory.[95] In fact, much of the Kew legacy quickly disappeared as the "Kew" work was dispersed into different departments and buildings in the large and expanding Teddington site. The testing of meteorological instruments, clinical thermometers, and watches was absorbed into the NPL's metrology division, while sextants, binoculars, and telescopes were tested in the optical division.[96]

At the same time, there was a similar dispersal of many members of the staff who had worked for so many years at Kew. Two of the observatory's longest-serving members of staff retired when the testing moved to Teddington. Thomas Baker, long the chief assistant under George Whipple and then Charles Chree, retired in September 1912. He had started at Kew in 1860, in its mid-Victorian heyday as a multifunctional observatory, just a year after Balfour Stewart had become superintendent. James Foster retired at the very end of 1912, after forty-six years of service at Kew. He spent the last two months of his career at Teddington helping to set up the barometer-testing operation there. Both men were awarded NPL pensions: Baker received £160 per annum, Foster £140.[97] As per the December 1909 proposal to the Treasury, the other senior members of the Kew staff remained on the NPL payroll and accompanied the standardization work when it moved to Teddington. For example, E. G. Constable, who had been in charge of the watch- and chronometer-rating operation at Kew, was still listed as a member of the scientific staff at Teddington in 1917. The four scientific staff who remained at Kew after 1910 to work for the Meteorological Office were younger and seemingly destined to follow long careers there. Yet only one of them, E. Boxall, appears on a group photograph of the Kew staff taken in 1927.[98]

Charles Chree must have been greatly relieved when the time-consuming standardization work moved away from Kew over 1912 and 1913. Chree was an efficient administrator, as is evidenced by the many volumes of his outgoing correspondence in the observatory archives. Yet Chree was first and foremost a researcher and a theoretician. Even in 1893, when applying for the position of superintendent, he had anxiously asked Robert Henry Scott for reassurance "that a considerable portion of the work expected from the Superintendent" would be research and not "routine work of the kind that would fall to the manager of a financial business."[99] Now, with much of the "routine work" gone, Chree would have more time for his research. He retired as superintendent in 1925, having served under three organizations: the Kew Committee, the NPL, and the Meteorological Office. All his successors were predominantly meteorologists who came to the superintendent's post after long careers at the Meteorological Office or in the imperial meteorological services scattered across the British Empire. Chree's immediate successor, Francis John Welsh Whipple, the younger son of Kew superintendent George Whipple, had worked for the office since 1912. Whipple's successor, James Martin Stagg, was coauthor with Chree of an important paper correlating solar activity with terrestrial magnetic disturbances. Yet Stagg is now best known for his weather forecast for 6 June 1944, which was pivotal to the launch of the D-Day assault on the Normandy beaches by Allied forces, an operation central to the liberation of Nazi-occupied Europe.[100]

Some magnetic instruments remained at Kew until 1926, when the last of them were transferred to Eskdalemuir. Seismology continued at Kew until well into the twentieth century: a Milne seismograph was still in use in the basement well after the First World War, and the observatory still had a seismograph in 1967. Today, the International Seismological Centre in Thatcham, England, has custody of the seismological data obtained at Kew from 1905 to 1968.[101] But for the most part, Kew Observatory after 1912 was simply a meteorological observatory. For the first time in its history, it specialized in just one science: even during the reign of George III, both astronomical and meteorological measurements had been carried out in the King's Observatory. Some major research projects were carried out at Kew after 1912. From the 1920s onward, innovative measurements of local air pollution were made at Kew. In 1941, researchers at Kew developed an improved version of the newly invented radiosonde, which sent back measurements of the upper atmosphere from an unmanned balloon. The radiosonde work strongly echoed the spirit of John Welsh and the Kew Committee in the 1850s, whose daring manned balloon ascents had made the first serious scientific explorations of the upper atmosphere. It also echoed the observatory's pre-1912 role as an instrument-testing center, since throughout the rest of the war years until 1946, Kew was a center for testing radiosondes. In the middle years of the twentieth century, Kew was a national center for the study of solar radiation, coordinating the results from other observatories.[102] A full history of the activities at Kew after 1912 remains to be written.

Yet, as had happened with the magnetic and meteorological work from the mid-1870s onward, these researches were not directed from Kew but carried out there on behalf of a client elsewhere. From 1912 on, that client was almost invariably the Meteorological Office. No longer was the research at Kew directed by a dedicated Kew Committee composed of private gentlemen, scientific servicemen, and, later on, university physicists. Kew in the twentieth century was not a center from which research was directed, the way it had been in the middle decades of the nineteenth century, nor a central testing laboratory, as it had been until the beginning of the twentieth. It was now reduced to functioning as a mere station of the Meteorological Office. Beginning in 1910, reports of the work at Kew were subsumed into an annual publication that included results from several Meteorological Office stations; in 1922, this became *The Observatories' Year Book*, which was published annually until 1967. The reports in these volumes were mostly tables of meteorological data, with only brief descriptions of the instruments. Gone were the detailed reports of the Kew Committee under BAAS and the Royal Society, complete with financial accounts, staff lists, and summaries of work done, which added up to more rounded accounts of the observatory's activities. Even in the 1920s, descriptions of the observatory suggested that an era had now ended. In 1920, the young meteorologist Frederick John Scrase noted a faded elegance about the building: "The place had more the air of a rather musty museum with a large number of instruments of historic interest."[103] A report in the 1922 *Observatories' Year Book* described the comparison of a barometer with two large standard mercury barometers at Kew, "which used to be the ultimate standards for the whole country when the testing of barometers was done at Kew Observatory."[104]

Chree's successors never forgot the legacy of the nineteenth-century Kew Observatory—in fact, they remained proud of it. The middle names of Francis John Welsh Whipple—son of superintendent George Whipple and solar observer Elizabeth Whipple (née Beckley)—clearly recall John Welsh, superintendent from 1852 to 1859. Francis's older brother, Robert Stewart Whipple, had a middle name recalling Welsh's successor, Balfour Stewart. In 1952, the elderly Robert Whipple dedicated a tablet in the superintendent's room at Kew to his grandfather, Robert Beckley, the mechanical assistant from 1853 to 1872.[105] In 1969, the Meteorological Office's *Meteorological Magazine*, descendant of the Victorian *Symons's Monthly Meteorological Magazine*, devoted an entire issue to the history of Kew Observatory—as told largely from the meteorologist's point of view.[106] As late as 2012, Malcolm Walker conferred upon Kew a revered status as a place where "some of the greatest names in the history of physics and meteorology" had worked and where Walker himself fondly remembered working in the 1960s when he was a student.[107]

Yet even by the time of the 1969 bicentenary, meteorologists were admitting to a feeling that the observatory's greatest days were over.[108] The observatory's specialization in just one branch of science, coupled with the fact that it was no longer a nerve center for observations or instrument standardization, meant that it was much less indispensable than it had been before 1912 and hence much more vulnerable to closure when Margaret Thatcher's government introduced severe funding cutbacks in 1980. To the consternation of some in the meteorological community, in 1980 the Ministry of Defence, by then the government body responsible for the Meteorological Office, made the decision to close Kew Observatory, which ceased operations as a meteorological observatory at the end of that year.[109] The observatory building remained Crown property. Until 2011, it was leased to a private company and used as an office building. At the time of this writing (2017), it is in the hands of a property developer.

The Royal Society's Gassiot Committee eventually outlived Kew Observatory as a scientific institution. By 1912, it had set up six subcommittees to direct specialized aspects of research at Kew and Eskdalemuir, such as seismology.[110] Thirty years later, at the height of the Second World War, its brief was broadened further by an agreement between the Royal Society and the Air Ministry (of which the Meteorological Office had been a part since 1920). Under this agreement, the Gassiot Committee became responsible for developing some technical and theoretical areas of meteorological research. For this purpose, the Royal Society Council supported the committee with grants of between £5,000 and £7,000— wholly different sums of money from the approximately £450 still coming in as dividends from the Gassiot Trust, an amount whose significance was much reduced by inflation in Great Britain after the First World War. The Gassiot Committee appointed three new subcommittees dealing with atmospheric composition and motion, atmospheric photochemistry and atmospheric radiation, and temperature. These subcommittees sponsored research projects in various universities. One project with great long-term significance was Gordon Dobson's Oxford-based research into atmospheric ozone. In 1985, scientists used one of Dobson's spectrophotometers to discover the notorious "hole" in the ozone layer above Antarctica. However, the Gassiot Committee activity with the highest profile in the mid-twentieth century began in 1953, when it initiated research into the upper atmosphere using rockets. This led to the Royal Society receiving a grant of £50,000 from the Air Ministry, with which a separate subcommittee coordinated work with British-built "Skylark" rockets. In the 1960s, this became part of the United Kingdom Space Research Programme, so effectively the Gassiot Committee's initiative helped to start space research in Britain. Some of the instruments placed aboard the rockets were designed and built by a team working at Kew.[111]

In the same decade, many of the Gassiot Committee's research programs were taken over directly by the Meteorological Office, and the committee was again reconstituted in 1967 with a more general brief "to stimulate meteorological research with special reference to the physical and chemical properties of the atmosphere." At the same time, the Royal Society used the income from the Gassiot Trust to help pay for one of two "Gassiot fellowships," with the one sponsored by the Gassiot Trust to be held at Kew Observatory. After the closure of Kew Observatory in 1980, the Royal Society paid the income of the trust fund to the University of Oxford to support the meteorological observations still being made on the site of the former Radcliffe Observatory.[112] Most strikingly, long before its closure, Kew was just a tiny part of the Gassiot Committee's wide range of activities. This underlined the reduced status of Kew Observatory generally: it was just one institution among the many spawned by the huge expansion of science in twentieth-century Britain.

CONCLUSION

William Napier Shaw's 1909 description of the recent opening of Eskdalemuir Observatory as "an epoch in the history of Kew" might be a fair description of the entire period between the establishment of the NPL in 1900 and the transfer of Kew to the Meteorological Office in 1910.[113] It was in these years that the observatory finally dropped its Victorian heritage and became a new kind of organization, more characteristic of the twentieth century than the nineteenth: a government scientific establishment, dedicated to just one branch of science.

Until the standardization work finally moved to Teddington in 1912 and 1913, Kew Observatory retained some of its pre-1900 technical aspects. Many of the staff who had worked at Kew since well before the turn of the century remained there until 1912. Throughout these years, the observatory also maintained its international status as a standardization center—to the point where there was a large market for thermometers bearing a forged Kew Observatory monogram. Nevertheless, after the NPL moved to Teddington in 1901, Kew's significance shrunk. It was now an isolated, outlying department of a larger organization. When the extension of London's electric tram network to Richmond compelled the NPL Executive Committee to move the Kew magnetic instruments to a new site in Scotland, the observatory's significance was reduced further in that the establishment of Eskdalemuir Observatory undermined Kew's raison d'être as a multifunctional observatory. Eskdalemuir was no mere outlying station of Kew: it was a "new" Kew in its own right, carrying out a full range of observations in meteorology and seismology, as well as geomagnetism, all directed by a physicist with the same academic background and scientific gravitas as Charles Chree, the superintendent of Kew.

The gentlemen's agreement between Shaw and Glazebrook to transfer responsibility for Kew in return for moving standardization to Teddington was clearly facilitated by the two men's almost identical backgrounds in Cavendish physics: from a scientific career point of view, they might have been twin brothers. Their shared Cavendish heritage meant that both were theoretically minded. Glazebrook would likely have sympathized with Shaw's attempt to change the culture of the Meteorological Office into that of a research-oriented organization, not one that emphasized the compilation of data, as it had been under Robert Henry Scott. This culture change meant that the office and its governing committee were less sympathetic than their predecessors to the Victorian conception of Kew as a nerve center for gathering data in several sciences. The observatory now had to fit into Shaw's new vision of meteorology, in which observation was subordinate to theory.

More significantly for the wider organization of science in the early twentieth century, the transfer of Kew Observatory from the NPL to the Meteorological Office in July 1910 led to it becoming a government observatory—at least, for the first time since the royal family relinquished it in 1841, prior to which it had been the monarch's personal property rather than a public institution reporting to a government department. With the 1910 transfer and especially the removal of instrument testing two years later, the last characteristics of the Victorian Kew Observatory disappeared. After 1912, it was no longer a mostly privately funded central institution for geomagnetism, meteorology, solar research, and instrument standardization. It was now a government station, specializing mostly in just one science and reporting to the Meteorological Office. Its staff and equipment, as well as the maintenance of the building, were all funded by government and not, as had mostly been the case before 1900, gentlemen-scientists and income from standardization.

The two-year delay in the transfer of standardization from Kew to Teddington can clearly be attributed to the Treasury's apparent unwillingness to pay for facilities at Teddington. Russell Moseley has attributed the Treasury's attitude toward the NPL in the first years of the twentieth century to its "incomprehension" of the value of scientific research.[114] Yet one might qualify Moseley's judgment after considering the individuals at the Treasury with whom the NPL and Royal Society dealt at the time of the standardization transfer. They principally corresponded with George Murray, the Treasury's sole permanent secretary since 1907, and Thomas Heath, assistant secretary from 1907 (and a permanent secretary from 1913). Murray came from a traditional upper-class background and had graduated from Oxford with third-class honors in classics, which makes it easy to infer that he had little knowledge of, or sympathy with, science or scientists. Heath, by contrast, was a grammar school–educated meritocrat who had excelled in both classics and mathematics at Cambridge. Since 1885, in his spare time, he had been publishing original scholarly works on ancient Greek mathematics; he is respected to this day as a major authority in the field. Heath had also served on the government's committee of enquiry that had reported in 1904 on the future of the Meteorological Office. There could have been few government officials in pre-1914 Britain more scientifically literate than Heath. Yet both men had a remarkably similar attitude toward spending public money. Murray, according to his biographer, inclined toward minimizing expenditure wherever possible and opposed the "supertax" brought in by Herbert Asquith's "new liberal" administration to pay for a radical expansion of the state's role in social welfare. Similarly, Heath showed a constant drive to minimize government spending, which reportedly led to friction between the Treasury and other government departments even before the "new liberal" welfare reforms began.[115] These welfare reform proposals were introduced at exactly the same time as the proposal to transfer Kew Observatory. The unprecedented increase in public spending that they implied was controversial enough to trigger a constitutional crisis, which led to the 1911 Parliament Act, restricting the power of Britain's House of Lords. It was against this background that Murray and Heath were being asked to find funding for what was, at the time, a hugely expensive plan to expand the facilities at the NPL.

The apparent reluctance of these Treasury officials to pay for new facilities at Teddington was not, therefore, necessarily because they were indifferent to the importance of science. Rather, it had more to do with the fact that they were opposed, in principle, to automatically spending large sums of public money, even on worthwhile causes, especially at a time of unprecedented demand on the public finances. Thus it was only when a private source offered half of the money required for the new facilities that the Treasury agreed, with a weary inevitability, to grant most of the rest of the funds over a series of installments over the next three years. The Treasury conceded that some expenditure was necessary—but only after other possible sources of funds had been tried. Murray and Heath were both steeped in the laissez-faire school of public finance, already encountered in, for example, the early 1890s decision to use Kew as the "nucleus" of the proposed national laboratory instead of building a completely new institution like the Berlin PTR. It was the Treasury's continued commitment to laissez-faire, not so much indifference to science, that the NPL, the Royal Society, and the Meteorological Office had to deal with right through to the removal of standardization from Kew to Teddington in 1912 and 1913. By then, an epoch had indeed passed in the history of Kew Observatory—but such an epoch had yet to pass in the history of the Treasury's economic policy.

CONCLUSION

The recording instruments [at Kew] designed by Stewart and Beckley have been in operation for nearly 70 years and in one way and another have played their part in teaching us about our climate. Long may they continue the good work.

F. J. W. WHIPPLE, 1937

Regular photographs of the sun would be—if only the practical problems could be solved—a quick means of building up an impersonal record of sunspot numbers and distribution. These problems were first solved at the Kew Observatory, which the British Association for the Advancement of Science supported near London.

KARL HUFBAUER, 1991

THE FIRST QUOTATION ABOVE IS THE CONCLUDING PARAGRAPH OF A presidential address given to the Royal Meteorological Society by Francis John Welsh Whipple, Charles Chree's successor as superintendent of Kew Observatory. To the historian, it offers an early illustration of how in the twentieth century, Kew Observatory was consecrated as an almost holy place for meteorologists, where "the good work" begun in the nineteenth century was carried on. The address bears the generic title of "Some aspects of the early history of Kew Observatory," yet perhaps naturally in a lecture to Britain's professional body of meteorologists, Whipple deals exclusively with the meteorological observations and instruments at Kew, making no mention of the geomagnetic, solar, and standardization work that was so important there in the nineteenth century and the beginning of the twentieth. On the other hand, the second epigraph, from a modern history of solar astronomy, mentions Kew only for its early contribution to solar physics, without acknowledging that solar photography was just one activity among many at Kew in the nineteenth century. Taken together, these two quotations dramatically illustrate how Kew Observatory has meant different things to people working in different branches of science. To the meteorologist, it was a pioneering meteorological observatory; to the astronomer, it was an early center for solar physics. This multiplicity of meanings attached to Kew Observatory might do much to explain why, so far, there has been no serious attempt to understand the *whole* of its work in the Victorian era and assess its overall historical significance. This book has attempted to do just that by writing its history as an institution and not just on isolated aspects of its work. This concluding chapter returns to the three great questions outlined in the introduction and shows how the findings presented in the rest of the book have helped to tackle them, before presenting some ideas for further research.

WHAT CAN THE HISTORY OF KEW OBSERVATORY TELL US ABOUT HOW THE PHYSICAL SCIENCES WERE ORGANIZED IN THE VICTORIAN ERA?

In the introduction, I divided this question into three subquestions. How were the physical sciences funded? How were they managed? What kind of people worked in these sciences? The history of Kew Observatory shows that there were some clear changes in the patronage of science between 1840 and 1910, yet the story also highlights an important similarity between the earlier and the later part of the period. In the early 1840s, Edward Sabine steered the project of a magnetic and meteorological observatory toward the privately funded British Association for the Advancement of Science. Sabine, an outsider in relation to the Cambridge network that dominated the physical sciences in Britain after 1815, was motivated by a determination to set up his own magnetic establishment, independent of the magnetic observatory erected at Greenwich by George Airy. The Astronomer Royal's astute moves in 1840—among them his stated wish to reduce the strain on the public purse—diverted what little government funding there was for geomagnetism and meteorology toward Greenwich, but that did not stop Sabine from establishing a scaled-down project at Kew with private funds. Thus in the 1840s, one man could manipulate the Royal Society and BAAS toward establishing an observatory, even without government money.

It is easy to think of the Royal Society government grant, introduced in 1850 by Lord Russell's government, as heralding the start of a new era in organized science. Even some contemporaries thought that it might be used to support the running of Kew Observatory. Yet it is debatable whether the grant in itself saved Kew from the attempts by BAAS in the late 1840s to close it down. The government intended the money for one-off projects by private individuals and not as an annually renewable grant to run larger institutions. Sabine and the Kew Committee were successful in obtaining government grant money, but only for individual projects. The Kew Committee obtained as much money from other sources, such as the Royal Society's private donation fund, as it did from government. The saving of Kew Observatory from closure was due more to the astuteness of Sabine and the businessman John Gassiot to commence instrument tests at Kew in return for fees. This standardization work was vital to keeping Kew running, not only through the income it generated but also because it made the case for closing the observatory much harder to argue—especially when the East India Company and then the Meteorological Department of the Board of Trade created a demand for large numbers of standardized instruments.

It could be argued that Kew was saved by the government in the 1850s, thanks to this demand from government departments. Yet it is important to remember that Kew remained independent of government. Its largest single source of income up to 1871 was the annual BAAS grant. It also did lucrative work for private individuals, instrument makers, and foreign governments. Although the Kew Observatory of the late 1850s was no longer entirely dependent on gentlemanly patronage, it was still a mostly privately funded organization. The successful 1860 expedition to Spain to photograph a solar eclipse tells a similar story: government support for this expedition was largely restricted to the loan of a ship, while equipment and personnel were primarily paid for by another wealthy businessman and devotee of science, Warren De La Rue. Similarly, funding for solar photography at Kew was entirely independent of government: its sources were BAAS, the Royal Society, and, again, De La Rue, who often paid assistants out of his own pocket. The narrative of solar photography at Kew also shows how it is difficult to categorize an observatory like Kew as "private" or "public." Kew in the 1860s was clearly not a "private" observatory of the kind found in a gentleman's private house or estate. Yet *public observatory* is a misleading term for an institution that was mostly, but not entirely, funded from private sources. Blanket categories such as "private" and "public" are insufficient in the case of Kew Observatory.

The transfer of Kew Observatory from BAAS to the Royal Society in 1871 accelerated a trend that began in the 1850s—the observatory became ever more dependent on standardization for income. Gentlemanly patronage, in the form of the Gassiot Trust, proved to be insufficient, especially after the trust's revenues dropped in the mid-1870s. So too was the even more paltry grant from the Meteorological Office: even as the office's "central observatory," Kew was little more dependent on government than before. By the time Robert Henry Scott wrote his oft-cited 1885 history of Kew Observatory, standardization was easily the largest source of the observatory's income. This trend continued after 1885. Thus by the time of Oliver Lodge's 1891 speech calling for a national laboratory like the prestigious new Physikalisch-Technische Reichsanstalt in Berlin, Kew was effectively a laboratory, the main function of which was testing instruments on a commercial basis. It was no longer a place for scientific investigations that it had predominantly been under BAAS. When the committee appointed in response to Lodge's speech was confronted with a lack of government support for a PTR-like institution, it was therefore an easy step for the committee to recommend a simple expansion of Kew Observatory, using it as the "nucleus" of a larger laboratory. Thus the origins of the NPL highlight an important similarity between the 1840s and the 1890s with regard to government patronage of science. The NPL came into existence in the way it did due to the sheer *lack* of government funding—in a way similar to when BAAS in the 1840s came to set up a scaled-down meteorological observatory, independent of government. The NPL was as much a product of laissez-faire as the Kew Observatory of the 1840s and 1850s, not a bold new departure by government. Even after 1900, the continuing dominance of a Victorian economic outlook does much to explain the Treasury's reluctance to fund the removal of standardization from Kew to Teddington. Exploring the NPL's origins through the evolution of Kew Observatory allows us to see more clearly the continuing importance of laissez-faire in the patronage of science throughout the Victorian era and into the early twentieth century.

The most striking change in the management of Kew Observatory over the 1840–1910 period is the gradual replacement of the gentlemen scientists who dominated the Kew Committee until the 1870s with university physicists, reflecting a change that took place in the overall management of the physical sciences in the Victorian era. The university physicists came to direct much of the research agenda, especially with regard to magnetic surveys and instrument standardization. The scientific servicemen remained well-represented on the Kew Committee after the 1870s, though none was ever as powerful as Sabine. However, aristocrats

and other independent men of wealth and influence remained important in obtaining patronage for science at Kew, as they did in government.
Francis Galton, the last of the gentlemen-scientists on the Kew Committee, remained its chairman right up to 1899 and was vital in the initial
liaison between BAAS and the Kew Committee that made the first move
to turn Kew into a national laboratory. The 1897 deputation to Lord
Salisbury that lobbied for the new laboratory was largely composed of
university physicists and engineers, yet it was led by Lord Rayleigh, a he-
reditary peer as well as a physicist who also had family connections to the
First Lord of the Treasury, Arthur Balfour. Indeed, this deputation bears
some resemblance to the party that visited Lord Melbourne in 1840 to
lobby for a "physical observatory": in both cases, well-connected, wealthy
men of science negotiated directly with aristocratic political leaders.

Some important changes in the kind of people who did science at Kew
took place between 1840 and 1900. Kew Observatory's first superintendent after 1842 was a gentleman volunteer, Francis Ronalds. By contrast,
all the superintendents from 1852 onward were paid employees of the
Kew Committee. Of these, all (except, as far as is known, Samuel Jeffery,
superintendent from 1871 to 1876) had received a university training in
the physical sciences. John Welsh and Balfour Stewart had both studied
at Edinburgh under James Forbes, but by the end of the century, the
position of superintendent was occupied by someone with a different
background. Charles Chree, although a Scot, had studied at the Cavendish Laboratory in Cambridge, by then a major training ground for
physicists. We need to be cautious, however, in reading the appointment
of Chree as indicative of the supremacy of the Cavendish at the end of
the nineteenth century. Of the four candidates short-listed for inter-
views in 1893, only two were from the Cavendish.[1] Chree's appointment
perhaps tells us more about the relative decline of the Edinburgh natural philosophical tradition and the larger number of university physics
departments and teaching laboratories by the 1890s. Edinburgh now
had competition from many other universities. The high standard of
applicants—one of whom was already FRS—also tells us something about
the prestige of Kew Observatory itself. The competition for the Kew
superintendentship in 1893 was as fierce as that for a university post at
the time.[2]

The scientific qualifications of the assistants at Kew became more
formalized toward the end of the nineteenth century. No longer were
the instrument readings being taken by a former sergeant from Sabine's

army regiment, as had been the case in the 1840s. Even the "boy clerks" had to have formal examination qualifications. Nevertheless, none of the regular assistants at Kew before 1910 were educated to degree level. Moreover, right up until 1910 these assistants remained chronically low-paid compared to their counterparts at Greenwich. This was a further continuity between Kew and the NPL: in the early years of the NPL at Teddington, well-qualified university physicists worked for salaries no higher than those of the senior assistants at Kew.[3] The cheapest—and most piecemeal—assistance of all was provided by women, such as Elizabeth Beckley, who took the solar photographs and helped to analyze the results. The story of Kew Observatory suggests that greater consideration needs to be given to the roles of lesser-known people in scientific practice. As the work of Elizabeth Beckley and John Welsh demonstrates, nineteenth-century scientific institutions, instruments, and discoveries were not straightforward realizations of the ideas of major figures such as John Herschel. They were much more complex and often reflected the social structure of Victorian England, in which credit often went to the grander figures at the expense of the lesser-known ones.

HOW DID THE "OBSERVATORY SCIENCES" AT KEW DEVELOP OVER THE COURSE OF THE NINETEENTH CENTURY?

Until the early 1870s, the observatory sciences at Kew showed a trend toward diversification. Geomagnetism, standardization, and solar physics were gradually added to the initial agenda of meteorology and atmospheric electricity. Astronomy was not central to Kew Observatory after 1842, the way it was at Greenwich, the university observatories, and the numerous private observatories operated by self-funded men of science. The closest it came to being central was in the late 1860s, when Balfour Stewart's theory-driven solar physics became key to his "cosmical physics" that connected solar activity with terrestrial magnetism and meteorology via the putative all-pervading ether. Yet this cosmical physics was frustrated partly by Sabine's insistence that Stewart devote his time to data collection and also by the vastly increased workload at Kew when it became the central observatory of the Meteorological Office.

With this multiplicity of activities, the diversification of the observatory sciences at Kew reached its peak in the late 1860s. After 1871, the transfer of the observatory to the Royal Society and the consequent lack of money reversed this trend by forcing Kew to become more specialized. Solar physics at Kew all but disappeared. Kew remained the Meteorological Office's central observatory that standardized the instruments for the office's other observatories, but from 1876 it no longer directed the observations. From then onward, the balance of power in this regard lay firmly with the Meteorological Office. Geomagnetism remained important at Kew, which was used as a base station for magnetic surveys and was still a place where other observatories—overseas as well as British—sent their instruments to be tested and their personnel to be trained in their use. Yet as with meteorology, Kew no longer led the work; it increasingly came to be directed by university physicists. Kew came to rely on the income from standardization, which became its specialty. The worse the financial situation became at Kew, the greater the range of instruments that began to be standardized there. Prominent among the list of instruments tested at Kew in the last three decades of the nineteenth century were clinical thermometers, instruments quite unconnected with the observatory sciences hitherto practiced at Kew yet which brought in essential income thanks to the lucrative medical instrument trade.

This move toward specialization at Kew in the last quarter of the nineteenth century mirrored a trend that also occurred at the Greenwich and Paris observatories at the same time. Mid-nineteenth-century Greenwich itself, with its own large magnetic and meteorological department in addition to its chronometer-testing service, had a diverse program of work in addition to astronomy. Yet meteorology was a relatively minor department at Greenwich by the end of the century. By then, too, the nerve center for French meteorological observations was a separate observatory—partly modeled on Kew—at Montsouris and not the old Paris Observatory, as had been the case when Le Verrier was director.

Kew did not just follow these trends. It was a major and active part of this changing dynamic in British national scientific institutions; indeed, events at Kew helped cause some of these changes, as the tussles between Airy and Kew in the 1870s demonstrate. Until the 1870s, for all Airy's dislike of Kew as a center for magnetic, meteorological, and solar observations, there was little he could do about it. Only with the changed financial situation at Kew after 1871—and after he had replaced Sabine as president of the Royal Society—did Airy have his chance to transfer the solar observations to Greenwich. By this time, however, even Airy was unable to seize the meteorological observations from Kew, as these were now firmly under the direction of the Meteorological Office. Neither Kew nor Greenwich could control the meteorological observations; a separate, specialist institution now oversaw them. Thus Kew became largely a specialist organization in its own right: by the end of the 1880s, it was effectively a national standardization laboratory. This specialization progressed further in the decade after 1900, when the magnetic observations moved to Eskdalemuir and standardization was transferred to Teddington, leaving Kew an almost purely meteorological institution. After 1910, different institutions specialized in just one of the observatory sciences carried out at Kew before 1871: Greenwich did astronomy (including solar physics), the Meteorological Office (of which Kew was a small part) ran the meteorological observations, and the NPL at Teddington became the undisputed center for instrument standardization.

We can draw two conclusions from this overall trend toward specialization at Kew. First, it was encouraged by the ideological climate of laissez-faire, which persisted despite all the controversy over the Devonshire Commission in the 1870s. After 1871, Kew needed to make money from standardization, as there was now no annual grant from BAAS and little government money. At the same time, Airy's moves to take possession of the Kew photoheliograph undercut the Devonshire Commission's proposal for a new, government-sponsored solar physics observatory that might rival Greenwich. Airy was thus able to manipulate laissez-faire to his advantage: by appearing to save the taxpayer money, he could stop his rivals from usurping his position as "national observer." David Aubin's assertion that the mid-nineteenth century was a time of crisis for observatories, in which they initially had to take on an increasing range of sciences before being forced to specialize (see the introduction), seems to be borne out by the experience at Kew. Yet he backs up his assertion with too much reliance on the French experience of state-run observatories. More room needs to be given to the British story, in which laissez-faire was central to the "crisis."

Secondly, the prominence of Kew Observatory in the overall move toward specialization in the various national institutions carrying out "observatory sciences"—astronomy at Greenwich, meteorology at the Meteorological Office, and standardization at Kew—means that Kew needs to be taken much more seriously when writing a narrative of the development of the observatory sciences in nineteenth-century Britain. The story is not simply about Greenwich. As Robert Smith has noted, more studies are needed that compare Greenwich with other observatories.[4] Perhaps more precisely, we could say that more studies are needed of Greenwich as part of the wider world of observatory sciences in the nineteenth century.

HOW DID STANDARDIZATION DEVELOP AT KEW IN THE CONTEXT OF
THE CULTURE OF THE PHYSICAL SCIENCES BETWEEN 1840 AND 1910?

By the late 1850s, standardization at Kew was an essential service—
to the London instrument trade and private devotees of science as well
as government departments. The establishment of the Meteorological
Department of the Board of Trade in 1854 certainly helped Kew, in the
sense that the department was a large and regular customer for standardized instruments to be used onboard ships. Yet Kew was also an essential
help to the Meteorological Department in that it provided the instruments needed for Robert FitzRoy's weather reports. In the same decade,
the standardization service also became important to observatories and
administrations across the British Empire; it was even recognized by
foreign observatories and governments. After 1871, the financial constraints facing Kew not only compelled it to rely on standardization as a
source of income: Kew, in turn, became indispensable to its expanding
customer base, which diversified into ever more types of instruments.
By the early 1890s, in addition to the Meteorological Office, other government departments, and the London instrument makers, clients for
Kew's services included the medical instrument trade and the watch and
chronometer industry. Moreover, the same service at Kew proved essential to the physics professors who managed the new university teaching and research laboratories that sprang up from the 1860s onward. By
the early 1890s, the prestigious Cavendish Laboratory and the related
Cambridge Scientific Instrument Company were clients for the service
at Kew.

Thus by 1891, when Oliver Lodge proposed a national laboratory for
testing instruments and determining standards, there was already a national standardization laboratory that did a substantial part of the work
of the institution that Lodge was calling for. This was admitted five years
later in the BAAS report on the proposed laboratory, which noted that
the work at Kew was of a type "strictly consistent with a large portion of
the work which would find a place in a national physical laboratory."[5]
Kew was the largest standardization laboratory in the British Isles and
the one that performed the widest range of work. Some work in electrical
standardization was done at the Cavendish Laboratory and in the Board
of Trade's electrical standards laboratory, but these places were small and
specialized. If anywhere was *the* place for general instrument standardization, it was Kew Observatory.

CONCLUSIONS AND SUGGESTIONS
FOR FURTHER RESEARCH

The findings presented in this book all suggest that Kew Observatory played an essential—and underrecognized—part in the culture of the physical sciences in the Victorian era. Also underrecognized is its importance in the origins of the National Physical Laboratory. Indeed, the most striking aspect of the story of Kew Observatory between 1840 and 1910 is the continuity between Kew and the NPL throughout the period. Right through the history of Kew from 1840 onward, we can trace early ideas for what we can now recognize as a kind of national standardization laboratory. Charles Wheatstone's 1842 prospectus for Kew suggested that it be used as a center for comparing instruments. In 1848, Francis Ronalds suggested that Kew could be taken over by the government as a "Proving House" for testing and comparing meteorological instruments. Even phrases similar to "national physical laboratory" had long been in existence by the 1890s. For example, the phrase "physical observatory" was first used by David Brewster in the 1830s, and in 1850 Roderick Murchison wrote to Herschel about Sabine's idea that a good use for the new Royal Society government grant might be to run "a good national Physical Observatory" at Kew. Even Lodge's 1891 speech only used phrases such as "physical laboratory" and "Physical Observatory": the exact phrase "National Physical Laboratory" was only first used a year later.

Thus when the NPL began work in 1900, Kew Observatory, the "nucleus" around which the NPL was built, already had a long history. The NPL evolved fully within the prevailing laissez-faire ideology. This did not immediately change after Kew became the NPL. The early NPL's low staff salaries—at both Kew and Teddington—and the expectation that it should be at least partly self-supporting through standardization fees[6] were a continuation of the same funding environment in which Kew had operated ever since the 1840s. Furthermore, even the apparently generous initial government spending on the NPL—£12,000 for the initial capital expenditure, plus an ongoing grant of £4,000 per annum thereafter—was not entirely without precedent: even in the 1870s, Greenwich Observatory was receiving nearly £7,000 per annum from the government, and the Meteorological Office, £10,000.[7] Viewed in this context, the "extension" of Kew Observatory in 1900 was modest indeed. With regard to the observatory sciences and standardization, laissez-faire persisted to 1914. Only after 1914, with the exigencies of total war and the subsequent formation of the government's Department of Scientific and Industrial Research, would this situation begin to change.[8] From the arguments presented in this book, we might assert that the establishment of the NPL at Kew Observatory in 1900 was not an early triumph of the state over laissez-faire but, rather, one of the last triumphs of the laissez-faire system in the patronage of science.

This book raises a host of further research questions related to Kew Observatory itself—for example, the interaction of Kew Observatory with the scientific instrument trade, something that this institutional history has only touched upon. Most exciting of all, however, are the broader issues in the historiography of the physical sciences that this book has highlighted. It is clear from the preceding chapters that historians need to take Kew Observatory much more seriously as a substantial, active component in the story of the physical sciences in nineteenth-century Britain. For example, the narrative of Kew's interactions with the wider world of astronomy and the physical sciences shows that there is a clear need for a more balanced historiography of astronomy and the other observatory sciences, one less heavily centered on Greenwich than has previously been the case. Secondly, the striking continuity between Kew Observatory and the NPL, both of which had to work within an unchanged laissez-faire political environment, contrasts starkly with the idea of the NPL as a break from the laissez-faire past and therefore suggests a need to look for more continuities between the science of the nineteenth century and that of the twentieth. The simplistic, presentist picture of a haphazard Victorian science being inevitably replaced by a twentieth-century model of government-supported science—such as that portrayed by Russell Moseley for the NPL[9]—needs to be critically reassessed and revised. Above all, the new insights presented in this book surely demonstrate the value of studying scientific institutions as a whole and not just individual aspects of them, as has hitherto been the case with Kew. More book-length histories of scientific institutions are needed that challenge existing assumptions about these institutions and about the history of science.

NOTES

INTRODUCTION: KEW OBSERVATORY, VICTORIAN SCIENCE, AND THE
"OBSERVATORY SCIENCES"

Epigraph: *BAAS Report* [1842], xxxv.

1. Walker, *History of the Meteorological Office*, 398; Galvin, "Kew Observatory."

2. Schaffer, "Where Experiments End"; Gooday, "Sunspots, Weather and the Unseen Universe."

3. Scott, "The History of the Kew Observatory"; Howarth, *The British Association for the Advancement of Science*, 154–69; Blackwell, "The Bicentenary of Kew Observatory," 235–38; Jacobs, "The 200-Years' Story of Kew Observatory."

4. Miller, "The Revival of the Physical Sciences in Britain"; see, esp., 107–19.

5. Sviedrys, "The Rise of Physics Laboratories in Britain"; Gooday, "Precision Measurement and the Genesis of Physics Teaching Laboratories in Victorian Britain."

6. Ross, "*Scientist*: The Story of a Word," 75–82.

7. Aubin, Bigg, and Sibum, *The Heavens on Earth*, 2–4; Aubin, "A History of Observatory Sciences and Techniques."

8. Aubin, "A History of Observatory Sciences and Techniques," 115–19.

9. Olesko, "Physics and Metrology"; see, esp., 713–14; Wise, "Introduction"; Schaffer, "Metrology, Metrication and Victorian Values," 440, 443; Crease, *World in the Balance*, 103.

10. Burton, "Robert FitzRoy and the Early History of the Meteorological Office"; Burton, "The History of the British Meteorological Office to 1905," 24–27; Walker, *History of the Meteorological Office*, 20–22.

11. Barrell, "Kew Observatory and the National Physical Laboratory," 171–74; Bryden, "Quality Control in the Making of Scientific Instruments."

12. Gooday, "Precision Measurement and the Genesis of Physics Teaching Laboratories in Victorian Britain" [1989]; Gooday, "Precision Measurement and the Genesis of Physics Teaching Laboratories in Victorian Britain" [1990]; Schaffer, "Late Victorian Metrology and its Instrumentation."

13. See, for example, Hall, *All Scientists Now*, 81–82 esp.; MacLeod, "Whigs and Savants."

14. Jeff Hughes has described an attempt by some Fellows in the 1930s to make the Royal Society more democratic. Hughes, "'Divine Right' or Democracy?"

15. Morrell and Thackray, *Gentlemen of Science*, 109–19.

16. MacLeod, 'The Royal Society and the Government Grant;" see, esp., 325–37.

17. Howarth, *The British Association for the Advancement of Science*, 151–54.

18. Cardwell, *The Organisation of Science in England*, 70; Alter, *The Reluctant Patron*, 138; Moseley, "Science, Government & Industrial Research," 1–15; Moseley, "The Origins and Early Years of the National Physical Laboratory," 222–24. The definition of "laissez-faire" broadly follows that in *OED*.

19. Strange, "On the Necessity for State Intervention to Secure the Progress of Physical Science."

20. Graeme Gooday has challenged Playfair's allegations of poor British performance in 1867, arguing that Playfair was selective in his assessment in order to strengthen his case for increased support for science education. Gooday, "Lies, Damned Lies and Declinism."

21. No book-length study of the Devonshire Commission has yet been published, but its origins and work are discussed in, for example, MacLeod, "The Support of Victorian Science," 202–7; MacLeod, "Science and the Treasury;" and Cardwell, *The Organisation of Science in England*, 119–26. Meadows, *Science and Controversy*, 75–111, discusses the commission at some length as part of a biographical study of its secretary, the astronomer Norman Lockyer. Lightman, "Huxley and the Devonshire Commission," assesses it in relation to the career of Thomas Huxley.

22. Cardwell, *The Organisation of Science in England*, 70.

23. MacLeod has noted that his own and his contemporaries' scholarly works of the 1970s "reveal the economic assumptions of the period in which they were written" and a belief in the inevitability of increased state intervention. MacLeod, "Introduction," viii–ix.

24. Cardwell, *The Organisation of Science in England*; see, esp., 177–78; Moseley, "Science, Government & Industrial Research," 39–66; Moseley, "The Origins and Early Years of the National Physical Laboratory"; Alter, *The Reluctant Patron*, 138–49; Pyatt, *The National Physical Laboratory*, 12–33; Magnello, *A Century of Measurement*, 11–30.

25. Moseley, "The Origins and Early Years of the National Physical Laboratory"; see, esp., 249.

26. Cannon, *Science in Culture*; Miller, "The Revival of the Physical Sciences in Britain, 1815-1840."

27. Herrmann, "An Exponential Law for the Establishment of Observatories in the Nineteenth Century"; also cited in Aubin, Bigg, and Sibum, *The Heavens on Earth*, 2; Aubin, "A History of Observatory Sciences and Techniques," III.

28. Dick, "National Observatories."

29. Smith, "A National Observatory Transformed"; Chapman, "Private Research and Public Duty"; Chapman, "Science and the Public Good"; Schaffer, "Astronomers Mark Time."

30. Hutchins, *British University Observatories 1772–1939*.

31. Lankford, "Amateurs and Astrophysics"; 275; Chapman, *The Victorian Amateur Astronomer*, xii.

32. Chapman, *The Victorian Amateur Astronomer*, 26.

33. Schaffer, "Where Experiments End: Tabletop Trials in Victorian Astronomy," 259.

34. Smith, "Remaking Astronomy."

35. Cawood, "Terrestrial Magnetism and the Development of International Collaboration in the Early Nineteenth Century."

36. Anderson, *Predicting the Weather*; see, esp., II–I2; Walker, *History of the Meteorological Office*, 9–I0; Wallace, "Meteorological Observation at the Radcliffe Observatory." Vladimir Jankovic has argued that there was widespread dissatisfaction with meteorology at the beginning of the nineteenth century because it had only recently shrugged off its Aristotelian heritage and lacked a sound theoretical basis. Jankovic, "The End of Classical Meteorology, *c.* I800"; see also Jankovic, *Reading the Skies*, 156–67.

37. Scott, "The History of the Kew Observatory," 37, 42; Rigaud, "Dr Demainbray and the King's Observatory at Kew," 282; Jacobs, "The 200-Years' Story of Kew Observatory," I62.

38. Beaufort to Herschel, 22 July 1839, RS:HS 3.407. References in this book to the correspondence of John Herschel and Edward Sabine follow the conventions used in Crowe, Dyck, and Kevin, *A Calendar of the Correspondence of Sir John Herschel*.

39. Jacobs, "The 200-Years' Story of Kew Observatory," I62, I63–I64; Morton and Wess, *Public & Private Science*, 29.

40. Rigaud, "Dr Demainbray and the King's Observatory at Kew," 282–83; Jacobs, "The 200-Years' Story of Kew Observatory," I65–66; Bugge, Moller Pedersen, and de Clercq, *An Observer of Observatories*, 77–81; Hutchins, "The King's Observatory, Richmond Park ('The Kew Observatory')," 2.

41. Rigaud, "Dr Demainbray and the King's Observatory at Kew," 283; Hutchins, "The King's Observatory, Richmond Park ('The Kew Observatory')," 3.

42. Bugge, Moller Pedersen, and de Clercq, *An Observer of Observatories*, 79–80.

43. Lindsay, "The Astronomical Instruments of HM King George III Presented to Armagh Observatory," 67; Bennett, *Church, State and Astronomy in Ireland*, 92–98.

44. Quill, *John Harrison*, 189–95.

45. Rigaud, "Dr Demainbray and the King's Observatory at Kew," 282.

46. Bugge, Moller Pedersen, and de Clercq, *An Observer of Observatories*, p. 81.

47. Ronalds, "Report Concerning the Observatory of the British Association at Kew, from August the 1st, 1843, to July the 31st, 1844," 121.

48. Cahan, "The Geopolitics and Architectural Design of a Metrological Laboratory"; Cahan, *An Institute for an Empire*.

49. Similarly, Rebekah Higgitt notes that the minutes of the Board of Visitors, Greenwich Observatory's governing body, "are frustratingly uninformative about discussion and disagreement among the Board's members." Higgitt, "A British National Observatory," 618.

50. Crowe, Dyck, and Kevin, *A Calendar of the Correspondence of Sir John Herschel*.

51. The largest of the other major repositories of Herschel letters is in the Harry Ransom Humanities Center at the University of Texas, from which scans of specific letters can be ordered online at minimal cost.

52. This exact phrase is used both in Walker, *History of the Meteorological Office*, 102, and by Jacobs, "The 200-Years' Story of Kew Observatory," 163.

CHAPTER 1: A "PHYSICAL OBSERVATORY": KEW, THE ROYAL SOCIETY, AND THE BRITISH ASSOCIATION, 1840–1845

An earlier version of part of this chapter was originally published in Macdonald, "Making Kew Observatory." I thank the editor and publisher of the *British Journal for the History of Science* for their permission to reuse the article in this book.

Epigraphs: Herschel, *Preliminary Discourse*, 213–14; Edward Sabine to John Herschel, 5 February 1841, RS:HS 15.123.

1. Miller, "The Revival of the Physical Sciences in Britain," 107–34, esp. 112–19.

2. Quoted in Walker, *History of the Meteorological Office*, 14.

3. Forbes, "Report upon the Recent Progress and Present State of Meteorology," 196. Part of this is quoted in Anderson, *Predicting the Weather*, 87–88, and Walker, *History of the Meteorological Office*, 15.

4. Forbes, "Supplementary Report on Meteorology," 144; Walker, *History of the Meteorological Office*, 15–16.

5. Cawood, "The Magnetic Crusade," 583–84.

6. Biographies of Sabine are all very brief. The most substantial are "Sir Edward Sabine"; Good, "Sir Edward Sabine"; and Reingold, "Sabine, Edward."

7. Some practitioners of science at this time, including Sabine, believed that each hemisphere might have two magnetic poles. See Enebakk, "Hansteen's Magnetometer and the Origin of the Magnetic Crusade," esp. 599.

8. Cawood, "The Magnetic Crusade," 502–7; Morrell and Thackray, *Gentlemen of Science* [1981], 356–59. Carter, "Magnetic Fever," 16–17, doubts whether Humboldt was really urged by Sabine to write to the Royal Society's president.

9. Cawood, "The Magnetic Crusade," 507; Buttmann, *The Shadow of the Telescope*, 121. Ruskin, *John Herschel's Cape Voyage*, 58–66, makes the case that Herschel's expedition led to him being appropriated as a hero of British imperialism as well as science.

10. The phrase "Magnetic Crusade" was not used during the lobbying for funding in the late 1830s. Contemporaries referred to it using language such as "fixed stations of observations" (Lord Minto to Lord Northampton, 7 January 1839, quoted in Morrell and Thackray, *Gentlemen of Science* [1981], 367). Carter, "Magnetic Fever," xv–xvi, suggests that the phrase was first used in 1842 in an American textbook on electricity and magnetism. More recently, Jessica Ratcliff, "Travancore's Magnetic Crusade," 326, has suggested that it was coined the same year by Humphrey Lloyd.

11. Brewster to Harcourt, 28 April 1832, in Morrell and Thackray, *Gentlemen of Science* [1984], 138–141; Harcourt to Brewster, 4 May 1832, in Morrell and Thackray, *Gentlemen of Science* [1984], 141.

12. Good, "A Shift of View," 55–56.

13. Herschel to Beaufort, 11 October 1835, RS:HS 21.188.

14. Ruskin, *John Herschel's Cape Voyage*, 52–57; Musselman, "Swords into Ploughshares."

15. Airy, *Autobiography of Sir George Biddell Airy*, 126, 131, 133, and 139.

16. Airy, *Autobiography*, 34–136; Cotter, "George Biddell Airy and his Mechanical Correction of the Magnetic Compass."

17. Miller, "The Revival of the Physical Sciences in Britain," 110; Aubin, "A History of Observatory Sciences and Techniques," 117. The importance of Airy's Cambridge education is described in Warwick, *Masters of Theory*, 72–75.

18. Morrell and Thackray, *Gentlemen of Science* [1981], 364.

19. Airy to W. S. Stratford, 27 February 1840, RGO 6.675,226.

20. Reingold, "Sabine, Edward," 51.

21. Babbage, *Reflections on the Decline of Science in England*, 76–101; Hall, *All Scientists Now*, 47–49.

22. Meadows, *Greenwich Observatory*, 103.

23. *Astronomical and Magnetical and Meteorological Observations made at the Royal Observatory, Greenwich, in the year 1839*, 95–100.

24. Hall, *All Scientists Now*, 155–56. Morrell and Thackray, *Gentlemen of Science* [1981], 350, refer to it as "a gaffe of the first order." Carter, "Magnetic Fever," 108–13, overemphasizes the role of John Herschel while attaching too little importance to that of Edward Sabine and overlooking evidence that the Royal Society withdrew its application to the government after the intervention of George Airy.

25. RS:CMB.284, 4 June 1840; RS:CMB.284: Report of Sub-Committee of Meteorology, 8 June 1840; Forbes to [Herschel], 4 June 1840, RS:MM.16.137; Forbes to [Herschel], 11 June 1840, RS:MM.16.138.

26. RS:CMB.284, 17 June 1840.

27. RS:CM, 18 June 1840.

28. Sheepshanks to Airy, 17 June 1840, RGO 6.675.228.

29. Lubbock to Herschel, 21 June 1840, RS:MM 16.141.

30. Lubbock to Herschel, 27 June 1840, RS:MM 16.142.

31. Sabine to Herschel, 6 July 1840, TNA:BJ 3/26; RS:CM, 18 June 1840.

32. Airy to Forbes, 8 October 1840, RGO 6.675.107.

33. Airy to Lord Northampton, 28 June 1840, RS:MM 11.145.

34. Sabine to Herschel, 18 June 1840, RS:MM 16.139. However, according to Herschel, Sabine suggested that a temporary observatory might operate at Woolwich until a permanent establishment was completed. Herschel to Airy, 6 July 1840, RGO 6.675.239.

35. Forbes to [Herschel], 4 June 1840, RS:MM.16.137.

36. Dawes to Herschel, 1 April 1839, RS:HS 6.58; Dawes to Airy, 1 April 1839, RGO 6.245.61.

37. RS:CM, 9 July 1840.

38. Morrell and Thackray, *Gentlemen of Science* [1981], 350.

39. Northampton to Lord Melbourne, RS:MM 16.145 (undated, but enclosed with a letter from Roberton to Herschel, 20 July 1840, RS:MM 16.144).

40. Airy, *Autobiography*, 142; Airy, *Report of the Astronomer Royal to the Board of Visitors*, 1841, 6.

41. Herschel to Airy, 6 July 1840, RGO 6.675.239.

42. Sabine to Herschel, 5 February 1841, RS:HS 15.123.

43. Beaufort to Herschel, 17 July 1839, RS:HS 3.40.

44. RS:CMB.284, 24 June 1841; RS:CM, 24 June 1841.

45. RS:CM, 11 November 1841.

46. RS:CM, 10 February 1842. The report is reproduced verbatim in Scott,

"The History of the Kew Observatory," 48–49. The original manuscript report is in RS:MM 16.189.

47. Sabine to Herschel, 13 January 1842, RS:HS 15.136.

48. Herschel to Sabine, 2 December 1841, TNA:BJ 3/26.

49. Herschel to Murchison, 15 February 1850, TxU:H/L-0269.

50. Herschel to Sabine, 5 September 1841, TNA:BJ 3/26.

51. RS:CM, 10 March 1842.

52. BAAS:CM, 28 March 1842.

53. BAAS 1842 "prospectus" for Kew Observatory, reprinted in Scott, "The History of the Kew Observatory," 50–52.

54. BAAS:CM, 28 March 1842; Earl of Lincoln to Lord Francis Egerton (BAAS), 26 May 1842, reproduced in BAAS:CM, 2 June 1842.

55. *BAAS Report*, 1842, xxii.

56. BAAS:CM, 25 September 1844; BAAS:CM, 17 June 1845; *BAAS Report*, 1845, xviii.

57. Murchison to Herschel, 16 June 1842, RS:HS 12.385.

58. Herschel to Wheatstone, 17 June 1842, TNA:BJ 3/26.

59. Herschel to Wheatstone, 17 June 1842, TNA:BJ 3/26.

60. BAAS:CM, 14 July 1842.

61. Ronalds to Samuel Carter, 21 February 1860, University College London Archives, GB 0103 MS ADD 206. I am grateful to Beverley Ronalds for this source. The most detailed biography of Ronalds is Beverley F. Ronalds, *Sir Francis Ronalds*; see, esp., 1–3 for Ronalds's background. See also Symons, "Ronalds, Sir Francis (1788–1873)."

62. BAAS:CM, 12 January 1843.

63. Ronalds, "Report concerning the Observatory of the British Association at Kew, from August the 1st, 1843, to July the 31st, 1844," 129.

64. Ronalds to Wheatstone, 16 November 1842, IET S.C.Mss.1/4/17b.

65. Ronalds, "Report concerning the Observatory of the British Association at Kew, from August the 1st, 1843, to July the 31st, 1844," 131.

66. Middleton, *A History of the Thermometer*, 41–42.

67. Middleton, *The History of the Barometer*, 319–20.

68. BAAS:CM, 17 June 1845.

69. Anderson, *Predicting the Weather*, 92–93.

70. Ronalds, "Report concerning the Observatory of the British Association at Kew, from August the 1st, 1843, to July the 31st, 1844," 121–26 and 130–31.

71. BAAS:CM, 1 December 1843.

72. Scott, "The History of the Kew Observatory," 51.

73. *BAAS Report*, 1843, xxxix.

74. Ronalds to Wheatstone, 16 November 1842, IET S.C.Mss.1/4/17b.

75. Ronalds, "Report concerning the Observatory of the British Association at Kew, from August the 1st, 1843, to July the 31st, 1844," 130–31.

76. Wheatstone to Sabine, 24 June 1842, RS:Sa. 1779.

77. Cawood, "The Magnetic Crusade," 515–16.

78. Sabine to William Radcliff Birt, 25 May 1848, RS:Sa. 1176.

79. Reingold, "Sabine, Edward," 51.

CHAPTER 2: SURVIVAL AND EXPANSION: KEW OBSERVATORY, THE
GOVERNMENT GRANT, AND STANDARDIZATION, 1845–1859

Epigraph: BAAS:CM, 31 January 1852.

1. Aubin, "A History of Observatory Sciences and Techniques," 115–19.

2. *BAAS Report* [1845], xviii.

3. "British Association for the Advancement of Science," *Times* (London), 28 June 1845, 8.

4. Orange, "The Beginnings of the British Association, 1831–1851," 58–59; Howarth, *The British Association for the Advancement of Science*, 35–36.

5. "Fifteenth Meeting of the British Association for the Advancement of Science," *The Athenaeum*, no. 922 (28 June 1845): 639–45.

6. Warwick, *Masters of Theory*, 72–75.

7. Astronomer Royal's Journal, 11 March 1842, RGO 6.24.

8. BAAS:CM, 16 January 1846.

9. Wheatstone to Herschel, 30 April 1846, RS:HS 18.151.

10. Report, dated "Kew Observatory, May 7, 1846," as reprinted in Scott, "The History of the Kew Observatory," 54–55. The report is also printed verbatim in 8 May 1846, BAAS:CM.

11. Herschel to Sabine, 20 April 1846, TNA:BJ 3/26. Wheatstone's report is also mentioned in Herschel to Airy, 29 April 1846, RGO 6.401.193.

12. Report, dated "Kew Observatory, May 7, 1846," as reprinted in Scott, "The History of the Kew Observatory," pp. 54–55.

13. BAAS:CM, 8 May 1846; *BAAS Report* [1846], xx.

14. Ronalds to Sabine, 30 June 1848, RS:Sa. 1087.

15. Among the very large literature on the "Neptune scandal," of particular relevance to the pressures on Airy in 1846 is Chapman, "Private Research and Public Duty," 124–26; and Smith, "The Cambridge Network in Action," 401 and 410–11.

16. BAAS:CM, 14 April 1848.

17. Herschel to Birt, 13 June 1842, RS:HS 19.96.

18. Jankovic, "Ideological Crests versus Empirical Troughs," esp. 34–39.

19. Herschel to Birt, 1 August 1849, RS:HS 19.138.

20. Birt to Herschel, 24 May 1848 (copy), RS:Sa. 164.

21. Herschel to Sabine, 1 June 1848, RS:Sa. 656.

22. Sabine to Herschel, 3 July 1848, RS:HS 15.224.

23. Sabine to Herschel, 17 June 1848, RS:HS 15.222.

24. Ronalds to Sabine, 30 June 1848, RS:Sa. 1087.

25. For short biographies of Birt, see Hutchins, "Birt, William Radcliff (1804–1881)"; "William Radcliff Birt" [1882a]; "William Radcliff Birt" [1882b].

26. Sabine to Herschel, 3 July 1848, RS:HS 15.224; Birt to Sabine, 4 July 1848, RS:Sa. 168.

27. Ronalds to Sabine, 30 June 1848, RS:Sa. 1087.

28. BAAS:CM, 7 July 1848.

29. Sabine to Herschel, 23 June 1848, RS:HS 15.223.

30. Herschel to Sabine, 21 July 1848, TNA:BJ 3/26.

31. Sabine to Herschel, 25 July 1848, TxU: 32.20–23 (M0523.6).

32. Herschel to Sabine, 31 July 1848, RS:Sa. 657.

33. Cawood, "The Magnetic Crusade," 514–15.

34. BAAS:CM, 9 August 1848.

35. Resolutions passed by BAAS General Committee, 16 August 1848, in BAAS:CM, 12 January 1849.

36. BAAS:CM, 12 January 1849.

37. Resolutions passed by BAAS General Committee at Birmingham on 19 September 1849, in BAAS:CM, 25 October 1849.

38. Northampton to Herschel, 18 September 1849, RS:HS. 5.277.

39. BAAS:CM, 25 October 1849.

40. Herschel to Sabine, 16 August 1849, TNA:BJ 3/84.

41. Birt to Professor [John] Phillips, 15 November 1849, RS:Sa. 169.

42. Birt to Herschel, 15 January 1850, RS:HS 4.137.

43. Eliza Birt (Birt's daughter) to Elisabeth Sabine (Sabine's wife), 28 December 1849, RS:Sa. 158; Ronalds to Sabine, 22 March 1850, RS:Sa. 1091.

44. Minutes of Kew Observatory Committee, 22 and 25 March 1850, in BAAS:CM, 9 April 1850.

45. Birt to Sabine, 5 June 1850, RS:Sa. 179.

46. Birt to Herschel, 14 June 1850, TxU: 29.20 (M0103); Herschel to Birt, 16 June 1850, TxU: 24.6 (L0100); Birt to Herschel, 19 June 1850, TxU: 29.20 (M0104). Birt's time at Kew is briefly described in Jankovic, "Ideological Crests versus Empirical Troughs," 37–38; and Ronalds, *Sir Francis Ronalds*, 352–56.

47. Phillips to [Forbes] Royle, 11 September 1850, RS:Sa. 995; Phillips to Royle, 25 September 1850, RS:Sa. 996.

48. Birt to Herschel, 14 June 1850, TxU: 29.20 (M0103).

49. Birt to Herschel, 21 December 1849, RS:HS 4.135.

50. Birt to Professor [John] Phillips, 15 November 1849, RS:Sa. 169.

51. Hartog (also McConnell), "Welsh, John (1824–1859)."

52. Schaffer, "Keeping the Books at Paramatta Observatory," esp. 120–25; Sweetman and McConnell, "Brisbane, Sir Thomas Makdougall, baronet (1773–1860)."

53. "Mr. John Welsh," xxxiv.

54. KCM, 5 July 1850.

55. BAAS:CM, 29 November 1852.

56. In August 1852, he expressed great anxiety to wind up his financial transactions in connection with Kew and "great need of relief from matters of this kind" due to illness and family matters. Ronalds to Sabine, 9 August 1852, RS:Sa. 1093. See also Ronalds, *Sir Francis Ronalds*, 363–64.

57. Ronalds to T. Romney Robinson, 18 February 1856, quoted in Ronalds, *Sir Francis Ronalds*, 365.

58. BAAS:CM, 25 October 1849.

59. BAAS:CM, 31 July 1850.

60. KCM, 5 July 1850.

61. KCM, 16 September 1850; Kew Diary, 27 August 1850–31 October 1851 (hereafter cited as KD).

62. Buttmann, *The Shadow of the Telescope*, 178.

63. BAAS:CM, 29 November 1850.

64. Ronalds to Sabine, 23 August 1850, RS:Sa. 1092.

65. *BAAS Report* [1859], lii.

66. Airy to Sykes, 16 August 1850, RGO 6.403.26.

67. Sykes to Airy, 25 November 1850, RGO 6.403.76.

68. Ronalds to Sabine, 23 August 1850, RS:Sa. 1092.

69. MacLeod, "The Royal Society and the Government Grant," 330–31. See also Hall, *All Scientists Now*, 147–51 and 163–64.

70. Murchison to Herschel, 2 March 1850, RS:MM 16.128. For background to Murchison's attitudes, see Orange, "The Beginnings of the British Association, 1831–1851," 53–57.

71. Murchison to Sabine, 9 February 1850, RS:Sa. 911; Murchison to Herschel, 11 February 1850, RS:MM 16.126.

72. Herschel to Murchison, 15 February 1850, TxU: 26.11 (L0269).

73. Murchison to Herschel, 2 March 1850, RS:MM 16.128. The recommendations presented in the Government Grant Committee's report of 7 March 1850 are very similar to Herschel's views. See Hall, *All Scientists Now*, 147–48.

74. BAAS:CM, 29 November 1850.

75. BAAS:CM, 2 July 1851.

76. MacLeod, "The Royal Society and the Government Grant," 325.

77. Sir John Herschel's Diary, RS:MS.584.

78. BAAS:CM, 25 October 1849; MacLeod, "The Royal Society and the Government Grant," 325.

79. BAAS:CM, 29 November 1852.

80. Quoted in Scott, "The History of the Kew Observatory," 51. Barrell, "Kew Observatory and the National Physical Laboratory," 171, also acknowledges this 1842 objective of Kew Observatory.

81. Gassiot to Sykes, 19 September 1850, RS:Sa. 591.

82. Middleton, *A History of the Thermometer and Its Use in Meteorology*, 108; Chang, *Inventing Temperature*, 75–76 and 83–84.

83. BAAS:CM, 29 November 1850.

84. KD, 22 February 1851; KCM, 25 March 1851; BAAS:CM, 11 April 1851.

85. KD, 29 July 1851 and 20 October 1851.

86. Sabine to Ronalds, 14 October 1850, quoted in Beverley F. Ronalds, *Sir Francis Ronalds*, 341.

87. KD, 22 March 1851.

88. BAAS:CM, 11 April 1851; Sabine to Welsh, 5 April 1852, TNA:BJ 1/11; Welsh to Sabine, 17 December 1851, TNA:BJ 3/32/38–40. Lord Rosse was then president of the Royal Society, and a traditional duty of that office was to host soirées for Fellows after Royal Society meetings.

89. BAAS:CM, 31 January 1852.

90. BAAS:CM, 31 January 1852.

91. Fleming, *Meteorology in America, 1800–1870*, 37; KCM, 16 September 1850; BAAS:CM, 2 July 1851.

92. BAAS:CM, 5 March 1852; Sabine to Welsh, 4 March 1852, TNA:BJ 1/11.

93. BAAS:CM, 20 September 1854.

94. Pang, *Empire and the Sun*, 142.

95. BAAS:CM, 5 March 1852.

96. BAAS:CM, 7 September 1853.

97. BAAS:CM, 20 September 1854.

98. BAAS:CM, 27 June 1855.

99. Williams, *The Precision Makers*, 24–26; Read, "History of the Firm Negretti & Zambra," 8.

100. Williams, *The Precision Makers*, p. 21.

101. Sabine to Welsh, 14 August 1853, TNA:BJ 1/11; BAAS:CM, 7 September 1853.

102. BAAS:CM, 29 November 1852.

103. BAAS:CM, 20 September 1854; Middleton, *The History of the Barometer*, 164–65; *A Treatise on Meteorological Instruments*.

104. BAAS:CM, 20 September 1854; BAAS:CM, 27 June 1855.

105. Dick, *Sky and Ocean Joined*, 62–67.

106. Woodward, *The Age of Reform 1815–1870*, 204–5 and 304–7.

107. BAAS:CM, 6 August 1856.

108. BAAS:CM, 6 August 1856 and 1 December 1856.

109. Receipts for verifications were as follows: £141 5s for 1856–1857, £110 for 1857–1858, and £69 12s 9d for 1858–1859 (BAAS:CM, 26 August 1857, 22 September 1858, and 14 September 1859).

110. BAAS:CM, 26 August 1857.

111. KCM, 30 May 1853.

112. Anderson, *Predicting the Weather*, 92.

113. Anderson, *Predicting the Weather*, 95; Walker, *History of the Meteorological Office*, 36–37.

114. Sabine, although an army officer, had connections with the Admiralty dating back to the 1820s, when he had been an original member of a Royal Society committee set up to advise the Admiralty (see chapter 1.)

115. BAAS:CM, 29 November 1852.

116. Bennett, "George Biddell Airy and Horology," 269–70.

117. This reduced priority was mirrored at Greenwich: in his annual reports, Airy mentions the Greenwich electrical apparatus only briefly after the mid-1840s and not at all as of the 1851 report. See, for example, Airy, *Report of the Astronomer Royal to the Board of Visitors* [1851], 9.

118. BAAS:CM, 6 August 1856.

119. Ronalds to Sabine, 30 June 1848, RS:Sa. 1087; BAAS:CM, 29 November 1852; Beverley F. Ronalds, *Sir Francis Ronalds*, 385.

120. Ronalds, "On Photographic Self-registering Meteorological and Magnetical Instruments."

121. Airy, *Report of the Astronomer Royal to the Board of Visitors* [1847], 5; Airy, *Report of the Astronomer Royal to the Board of Visitors* [1848], 9–10.

122. Airy to Secretary to Admiralty, 17 August 1848, RGO 6.675.559; Herschel to Airy, 26 October 1848, RGO 6.675.581; Beverley F. Ronalds, *Sir Francis Ronalds*, 483–93.

123. John Herschel's Diary, 23 March 1849, RS:MS.584; BAAS:CM, 20 April 1849 and 12 September 1849; Northampton to Sabine, 27 April 1849, RS:Sa. 947.

124. KD, 1 April 1851; BAAS:CM, 11 April 1851.

125. BAAS:CM, 31 January 1852; BAAS:CM, 31 July 1850; BAAS:CM, 11 April 1851.

126. BAAS:CM, 9 April 1850.

127. Welsh to Sabine, 19 November 1854, RS:Sa. 1734.

128. BAAS:CM, 9 April 1850; KCM, 24 April 1858. The British government paid for the Toronto instrument.

129. BAAS:CM, 26 August 1857 and 26 April 1858.

130. Tombs, *France 1814–1914*, 188–89; Baguley, *Napoleon III and his Regime*, 194–95.

131. RS:CM, 14 December 1854; KCM, 23 January 1855.

132. BAAS:CM, 27 June 1855 and 6 August 1856; Welsh to Sabine, 7 January 1856, RS:Sa. 1737.

133. Beverley F. Ronalds, *Sir Francis Ronalds*, 369–70.

134. Ronalds, "On Photographic Self-Registering Meteorological and Magnetical Instruments."

135. BAAS:CM, 20 September 1854; Middleton, *A History of the Thermometer and Its Use in Meteorology*, 219–21.

136. Tucker, "Voyages of Discovery on Oceans of Air," esp. 146–48; Rolt, *The Aeronauts*, 190–92.

137. BAAS:CM, 5 March 1852 and 29 November 1852. Examples of the BAAS circulars can be found in the papers of George Airy—see, for example, RGO 6.402.457.

138. Welsh, "An Account of Meteorological Observations in Four Balloon Ascents, Made under the Direction of the Kew Observatory Committee of the British Association for the Advancement of Science," esp. 338–40.

139. "Scientific Balloon Ascent," *Illustrated London News*, no. 577 (4 September 1852). The picture is briefly mentioned in Tucker, "Voyages of Discovery on Oceans of Air," 150. The 17 August ascent is also briefly described in "Scientific Balloon Ascent," *Illustrated London News*, no. 576 (28 August 1852).

140. *BAAS Report* [1858], xxxix.

141. Rolt, *The Aeronauts*, 192–94.

142. Tucker, "Voyages of Discovery on Oceans of Air," 169–71.

143. Quotation in Anderson, *Predicting the Weather*, 99; "Mr. John Welsh," xxxvii.

144. KCM, 1 August 1859 and 23 May 1862. An engraving of the shelf of instruments used by Glaisher bears a striking resemblance to a photograph of the instruments used by Welsh in 1852. See Glaisher, *Travels in the Air*, 39, and Shaw, "An Episode in the History of Kew Observatory," frontispiece.

145. KCM, 5 July 1850; KD, 1 April 1851 and 21 June 1851; Welsh to Sabine, 3 March 1852, TNA:BJ 3/32; KCM, 7 March 1852. Nicklin lost his life at sea in 1854 while working as a photographer in the Crimean War. See Beverley F. Ronalds, *Sir Francis Ronalds*, 549–50.

146. KCM, 12 November 1853 and 3 December 1853; Beckley, "Description of a Self-recording Anemometer"; BAAS:CM, 27 June 1855.

147. KCM, 11 June 1855; Panwitz and Schwarz, *Alexander von Humboldt Familie Mendelssohn Briefwechsel*, 472; Welsh to Sabine, 7 January 1856, RS:Sa. 1737.

148. BAAS:CM, 6 August 1856.

149. KCM, 22 July 1856 and 7 November 1856. The employment of Chambers at Kew is also noted by Donald Cardwell in *The Organisation of Science in England*, 84–85.

150. KCM, 24 April 1858; BAAS:CM, 22 September 1858.

151. Chapman, "Science and the Public Good," 52.

152. Welsh to Forbes, 6 September 1856, St Andrews msdep / 7 / Incoming / 1856 / 85a and b.

153. John Welsh, Royal Society Election Certificate [1857], RS:EC 1857.14.

154. Sabine to Herschel, 6 April 1857, RS:HS 15.242.

155. *BAAS Report*, 1859, xlvi.

156. KCM, 13 September 1854; Anderson, *Predicting the Weather*, 88–89.

CHAPTER 3: "SOLAR SPOT MANIA," "COSMICAL PHYSICS," AND METEOROLOGY, 1852–1870

An earlier version of part of this chapter was originally published in Macdonald, "Solar Spot Mania." I thank the editor and publisher of the *Journal for the History of Astronomy* for their permission to reuse the article in this book.

Epigraphs: Welsh to Sabine, 23 April 1852, TNA:BJ 3/32; "The Meteorological Department," *The Athenaeum*, no. 2136 (3 Oct. 1868): 436–437.

1. This is approximately the story told in Hufbauer, *Exploring the Sun*, 46–67, and King, *The History of the Telescope*, 226.

2. Charbonneau, "The Rise and Fall of the First Solar Cycle Model"; Goo-

day, "Sunspots, Weather and the Unseen Universe"; Anderson, *Predicting the Weather*, 250–76; Meadows, *Science and controversy*, 122–28.

3. Hufbauer, *Exploring the Sun*, 35–40; Hoskin, *Discoverers of the Universe*, 148–49.

4. Hufbauer, *Exploring the Sun*, 46; Charbonneau, "The Rise and Fall of the First Solar Cycle Model," 351.

5. John and Lady Herschel to Caroline Herschel, 10 January 1837, in Evans et al., *Herschel at the Cape*, 280–82; Herschel, *Results of Astronomical Observations*, 431.

6. Herschel to Francis Baily, 1 March 1837, RS:HS 3.138; Meadows and Kennedy, "The Origin of Solar-Terrestrial Studies," 424—briefly note Herschel's 1837 speculation on a possible link between sunspots and aurorae.

7. Herschel, *Results of Astronomical Observations*, 432–45. "Kalotype" or "calotype" was the name given to William Henry Fox Talbot's second photographic process, invented in 1841 and an improvement on the "photogenic drawings" he introduced in the 1830s.

8. Herschel, "Extract of letter respecting Mr. Griesbach's communication on the solar spots."

9. Draft of letter from Herschel to Talbot, 30 January 1839, quoted in Schaaf, *Out of the Shadows*, 50.

10. Buttmann, *The Shadow of the Telescope*, 136–52; Schaaf, *Out of the Shadows*, 45–102.

11. Sabine to Herschel, 16 March 1852, RS:HS 15.235; Sabine to Herschel, 12 April 1852, RS:HS 15.236.

12. Herschel to Faraday, 10 November 1852, RS:HS 23.127.

13. Charbonneau, "The Rise and Fall of the First Solar Cycle Model," 355; Le Conte, "Warren De La Rue," 23; Rothermel, "Images of the Sun," 152.

14. Welsh to Sabine, 23 April 1852, TNA: BJ 3/32. The Kew Diary for 1850–1851 confirms that on 21 June 1851, Welsh and an assistant (Nicklin) "procured some Daguerreotype pictures of the sun with the help of a small reflecting telescope" (KD, under entry for 1851 June 27th).

15. Herschel to Gassiot, 24 April 1854, printed in *BAAS Report*, 1854, xxxiv–xxxv; Herschel, "On the Application of Photography to Astronomical Observations," 158–59. The RAS version is addressed to "Colonel Sabine," but both the BAAS version and the original manuscript (see below, note 18) are clearly addressed to Gassiot.

16. *BAAS Report* [1854], xxxiv.

17. KCM, 15 March 1854.

18. Herschel to Gassiot, 24 April 1854, RS:MC. 5.164.

19. Accounts of Herschel's life emphasize how he was overworked in these

years, to the point of a nervous breakdown later in 1854. See, for example, Buttmann, *The Shadow of the Telescope*, 176–83.

20. Crowe, Dyck, and Kevin, *A Calendar of the Correspondence of Sir John Herschel*.

21. Gassiot to Herschel, 9 May 1854, RS:HS 8.59; *BAAS Report* [1854], xxxv.

22. Schaffer, "Where Experiments End," 259.

23. Tobin, *The Life and Science of Léon Foucault*, 52–54.

24. Hartog (also Meadows), "Rue, Warren de la (1815–1889)"; K[nobel], "Warren De La Rue," 157–60; Le Conte, "Warren De La Rue," 16.

25. KCM, 15 March 1854; *BAAS Report* [1854], xxxv.

26. De La Rue, "Report on the present state of Celestial Photography in England," 149–53.

27. BAAS:CM, 27 June 1855; De La Rue to Herschel, 12 October 1856, RS:HS 6.137.

28. De La Rue to Herschel, 28 September 1857, RS:HS 6.140.

29. BAAS:CM, 14 September 1859.

30. BAAS:CM, 17 November 1859.

31. BAAS:CM, 27 June 1860.

32. BAAS:CM, 4 September 1861.

33. De La Rue to Herschel, 14 September 1862, RS:HS 6.150.

34. *BAAS Report* [1862], xxxvi; BAAS:CM, 22 November 1861.

35. De La Rue to Herschel, 14 September 1862, RS:HS 6.150.

36. *BAAS Report* [1863], xxxviii.

37. De La Rue et al., "A Comparison of the Kew Results of Observations on Sun-Spots," 77. The employment of Elizabeth Beckley is briefly mentioned in Schaffer, "Where Experiments End," p. 271, and Le Conte, "Warren De La Rue," 33.

38. KCM, 12 November 1853; Sabine to [Balfour Stewart], 6 July 1861, TNA: BJ 1/29; Sun Diary, 1863–1865, RGO 57/11.

39. Fara, *Pandora's Breeches*, 30–144 (on Elisabetha Hevelius), and 145–66 (on Caroline Herschel).

40. Cited in Becker, *Unravelling Starlight*, 182.

41. Anderson, *Predicting the Weather*, 280–81.

42. Rothermel, "Images of the Sun," 147–50; Hufbauer, *Exploring the Sun*, 44.

43. De La Rue, "The Bakerian Lecture," 333–34.

44. BAAS:CM, 27 June 1860.

45. De La Rue, "The Bakerian Lecture," 355; Airy, "Account of Observations of the Total Solar Eclipse of 1860, July 18," 2.

46. *BAAS Report* [1861], xxxv.

47. Pang, *Empire and the Sun*, 121–43. In the same work (11–48), Pang more generally describes the complexity of organizing eclipse expeditions in the Victorian era.

48. Airy, "Account of Observations of the Total Solar Eclipse of 1860, July 18," 2–3.

49. De La Rue, "The Bakerian Lecture," 407–15.

50. *BAAS Report* [1861], xxxv.

51. Sabine to Stewart, 7 July 1860, TNA:BJ 1/29.

52. *BAAS Report* [1861], xxxv.

53. Ratcliff, *The Transit of Venus Enterprise in Victorian Britain*, 17; Ashworth, "John Herschel, George Airy and the Roaming Eye of the State."

54. De La Rue, "The Bakerian Lecture," 333; De La Rue to Herschel, 14 September 1862, RS:HS 6.150; *BAAS Report* [1862], xxxvii.

55. *BAAS Report* [1866], xxxvi; *BAAS Report* [1870], xlviii.

56. Welsh to Sabine, 27 September 1858, RS:Sa. 1764; Welsh to Sabine, 6 December 1858, RS:Sa. 1767; Samuel J Fox to [headed "Copy; to Dr Bence Jones"], 27 December 1858, RS:Sa. 1768; "Mr. John Welsh," xxxiv–xxxviii.

57. KCM, 1 August 1859; BAAS:CM, 14 September 1859.

58. Anderson, *Predicting the Weather*, 145.

59. KCM, 1 August 1859; Clerke (also) Gross, "Breen, James (1826–1866)"; Hutchins, *British University Observatories 1772–1939*, 71 and 126.

60. Forbes to Welsh, 18 December 1855, TNA:BJ 1/9; Forbes to Welsh, 20 October 1855, TNA:BJ 1/9.

61. KCM, 1 August 1859.

62. The dispute between Stewart and Kirchhoff and their respective supporters is described in Siegel, "Balfour Stewart and Gustav Robert Kirchhoff." The most detailed biography of Stewart is Schuster, "Memoir of the late Professor Balfour Stewart." Other biographies can be found in T[ait], "Dr. Balfour Stewart"; Schuster, *Biographical Fragments*, 206–15; and Hartog (also Gooday), "Stewart, Balfour (1828–1887)."

63. Carrington, "Description of a Singular Appearance Seen in the Sun on September 1, 1859"; Hodgson, "On a Curious Appearance seen in the Sun"; Stewart to Sabine, 14 November 1859, RS:Sa. 1380.

64. The 1859 phenomenon is now known to have been a "white-light flare," an extremely rare type of solar flare that is intense enough to be seen in ordinary visible light.

65. Stewart, "On the Great Magnetic Disturbance of August 28 to September 7, 1859."

66. Sabine to Stewart, 26 November 1860, TNA:BJ 1/29.

67. Stewart to Sabine, 18 October 1861, RS:Sa. 1459; Sabine to Stewart, 21 October 1861, TNA:BJ 1/29.

68. Stewart to Sabine, 27 August 1862, RS:Sa. 1479; Sabine to Stewart, 31 August 1862, TNA:BJ 1/30; Stewart to Sabine, 1 September 1862, RS:Sa. 1481.

69. Smith, *The Science of Energy*, 63; Stewart, "On the Nature of Those Red Protuberances which Are Seen on the Sun's Limb during a Total Eclipse," esp. 304; Van Allen and Bagenal, "Planetary Magnetospheres and the Interplanetary Medium," 50.

70. Balfour Stewart, Royal Society Election Certificate [1862], RS:EC/1862/12.

71. Stewart to Sabine, 1 September 1862, RS:Sa. 1481.

72. Stewart to Sabine, 17 February 1864, RS:Sa. 1512; BAAS:CM, 14 September 1864; "Flagstaff Observatory, 1858–1863," http://museumvictoria.com.au/collections/themes/1630/flagstaff-observatory-1858–1863; Schuster, *Biographical Fragments*, 213–14.

73. Stewart, "On Sun-Spots and Their Connection with Planetary Configurations." For background on nineteenth-century ether theory, see Cantor and Hodge, *Conceptions of Ether*, 49–50.

74. Stewart to Sabine, 12 July 1864, RS:Sa. 1515.

75. Gooday, "Sunspots, Weather and the Unseen Universe," 130.

76. Stewart to Sabine, 29 December 1863, RS:Sa. 1510.

77. Stewart and Tait, "Preliminary Note on the Radiation from a Revolving Disc"; Stewart and Tait, "On the Heating of a Disc by Rapid Rotation in Vacuo."

78. Stewart to Sabine, 17 February 1865, RS:Sa. 1538.

79. Stewart to Sabine, 27 April 1865, RS:Sa. 1540; Stewart to Sabine, 13 May 1865, RS:Sa. 1544.

80. Smith, *The Science of Energy*, 253–55; Gooday, "Sunspots, Weather and the Unseen Universe," esp. 128.

81. Stewart and Lockyer, "The Sun as a Type of the Material Universe. Part II," esp. 327.

82. Meadows, *Science and Controversy*, 51–60.

83. Gooday, "Sunspots, Weather and the Unseen Universe," 119.

84. Sabine to Stewart, 6 January 1863, TNA:BJ 1/30; Stewart to Sabine, 8 January 1863, RS:Sa. 1496.

85. Stewart to Sabine, 16 January 1865, RS:Sa. 1535.

86. Stewart to Sabine, 13 May 1865, RS:Sa. 1544.

87. Stewart to Sabine, 7 December 1866, RS:Sa. 1580.

88. Gooday, "Precision Measurement and the Genesis of Physics Teaching Laboratories in Victorian Britain" (1989), 32–40, esp. 32; *BAAS Report* [1870], lvi; Smith, *The Science of Energy*, 178 and 179.

89. Sabine to T. H. Farrer (Board of Trade), 15 June 1865, in *Report of a Committee Appointed to Consider Certain Questions Relating to the Meteorological Department of the Board of Trade* (1866), iii–v.

90. Burton, "The History of the British Meteorological Office to 1905," 59 and 255; Walker, *History of the Meteorological Office*, 60.

91. *Report of a Committee Appointed to Consider Certain Questions Relating to the Meteorological Department of the Board of Trade* (1866).

92. Anderson, *Predicting the Weather*, 123–124.

93. BAAS:CM, 28 June 1861; Galton to Stewart, 5 July 1860, TNA:BJ 1/24. In his autobiography (Galton, *Memories of My Life*, 225), Galton notes Sabine's influence in bringing him onto the Kew Committee. The minutes of the Kew Committee (KCM, 28 May 1860) confirm that Galton was elected to the Kew Committee "on the proposition of Genl. Sabine."

94. Galton to Sabine, 16 March 1866, RS:Sa.5 86; also cited in Anderson, *Predicting the Weather*, 140, and Burton, "The History of the British Meteorological Office to 1905," 58.

95. Airy to George Gabriel Stokes, 31 March 1866, RS:MC. 7.317; RS:CM, 19 April 1866. Airy's letters to the council are briefly referred to in Anderson, *Predicting the Weather*, 143, and Walker, *History of the Meteorological Office*, 59–60.

96. "Address to the Visitors of the Royal Observatory, Greenwich, by the Astronomer Royal. (*Communicated Only to the Members of the Board of Visitors*)," January 1863, RGO 55. Above the heading of the document is the statement "*Confidential Document, Privately Printed.*"

97. Stewart to Gassiot, 17 May 1866, RS:Sa. 1571; Stewart to Gassiot, 28 May 1866, RS:Sa. 1572; Stewart to Gassiot, 7 June 1866, RS:Sa. 1573.

98. Stewart to Gassiot, 28 May 1866, RS:Sa. 1572; BAAS:CM, 4 September 1867.

99. Stewart to Sabine, 7 October 1861, RS:Sa. 1458.

100. BAAS:CM, 4 September 1867; Burton, "The History of the British Meteorological Office to 1905," 70–72; *BAAS Report* [1868], xliii and xliv.

101. Stewart to Sabine, 5 January 1867, RS:Sa. 1586; Gassiot to Stewart, 23 July 1867, TNA:BJ 1/24.

102. Sabine to Stewart, 31 July 1867, TNA:BJ 1/30.

103. Quoted in Gooday, "Precision Measurement and the Genesis of Physics Teaching Laboratories in Victorian Britain" (1989), 7–36.

104. *BAAS Report* [1869], xlv; BAAS:CM, 6 September 1865; Jackson & Co. (brewery) to Kew Observatory Secretary, 17 August 1869, TNA:BJ 1/26.

105. Davis, "Weather Forecasting and the Development of Meteorological Theory at the Paris Observatory, 1853–1878," 363 and 377–79.

106. BAAS:CM, 4 September 1867.

107. Gooday, "Sunspots, Weather and the Unseen Universe," 127.

108. "The Meteorological Department," *The Athenaeum*, no. 2136 (3 October 1868): 436–37.

109. Stewart, "Physical Meteorology. I. Its Present Position," 102.

110. "The Meteorological Committee," *The Saturday Review* 26, no. 680 (7 November 1868): 622–23.

111. Gassiot to Lockyer, 17 November 1868, TNA:BJ 1/24.

112. Stewart to Gassiot, 8 November 1868, RS:Sa. 1634.

113. Stewart to Sabine, 15 January 1869, RS:Sa. 1637.

114. Stewart to Sabine, 2 April 1869, RS:Sa. 1641.

115. Stewart to Gassiot, 8 October 1869, with RS:Sa. 1656; statement by Gassiot [October 1869?], RS:Sa. 1656.

116. Stewart to Gassiot, 13 October 1869, with RS:Sa. 1656.

117. Statement by Gassiot [October 1869?], RS:Sa. 1656; Schuster, *Biographical Fragments*, 210. Schuster does not name Chambers as the proposed successor to Stewart.

118. Sabine to Stewart, 31 May 1870, RS:Sa. 1661; also quoted in Schuster, *Biographical Fragments*, 210.

119. Stewart to Sabine, 8 July 1870, RS:Sa. 1663.

120. For an accurate popular account, see Clark, *The Sun Kings*.

121. *BAAS Report* [1870], xlviii.

122. *BAAS Report* [1870], lvii.

123. Schaffer, "Where Experiments End," 259.

CHAPTER 4: KEW OBSERVATORY AND THE ROYAL SOCIETY, 1869–1885

Epigraphs: Gassiot to Sharpey, 4 July 1871, RS:MS. 843.30; Galton, "Description of the Process of Verifying Thermometers at the Kew Observatory," 84.

1. Essentially this story is told in Scott, "The History of the Kew Observatory," 62–63; Jacobs, "The 200-Years' Story of Kew Observatory," 163; Hall, *All Scientists Now*, 186; and Walker, *History of the Meteorological Office*, 102–3.

2. MacLeod, "The Support of Victorian Science," 211.

3. *BAAS Report* [1869], xlv.

4. Hall, *All Scientists Now*, 106.

5. BAAS:CM, 13 November 1869.

6. KCM, 11 December 1869.

7. BAAS:CM, 11 December 1869.

8. *BAAS Report* [1870], xlv–xlvi.

9. BAAS:CM, 5 November 1870.

10. RS:CM, 15 December 1870, 19 January 1871, and 16 February 1871.

11. Gassiot to Herschel, 13 February 1871, RS:HS 8.67; Gassiot to Airy, 13 February 1871, RGO 6.394.

12. RS:CM, 16 March 1871. Thomson's reply, dated 20 March, is in RS:CM, 20 April 1871.

13. Stewart to Gassiot, 8 November 1870; Lloyd to Gassiot, 14 February 1871; Sykes to Gassiot, 14 February 1871; Robinson to Gassiot, 14 February 1871; Airy to Gassiot, 13 February 1871, all in RS:CM, 16 March 1871; Thomson to Gassiot, 20 March 1871, in RS:CM, 20 April 1871.

14. Herschel to Gassiot, 17 February 1871, in RS:CM, 16 March 1871.

15. Herschel to Gassiot, 17 February 1871, RS:HS 8.68.

16. Bartholomew, "The Discovery of the Solar Granulation," 284.

17. Herschel to [Whipple?], 19 February 1871, TNA:BJ 1/83; Herschel to Whipple, 1 March 1871, TNA:BJ 1/83.

18. Gassiot to Sharpey, 13 March 1871, RS:MC. 9.178; partly summarized in RS:CM, 16 March 1871.

19. Minutes of Kew Maintenance Committee, 28 March 1871, RS:CMB. 2.05.

20. RS:CM, 27 April and 25 May 1871.

21. Scott, "Copy of Statement Sent to Mr Gassiott [*sic*] May 22. 1871," TNA:BJ 1/92.

22. Gassiot to Walter White (assistant secretary, Royal Society), 23 May 1871, RS:MC. 9.205.

23. Thomson, "Address of Sir William Thomson, Knt., LL.D., F.R.S., President," lxxxv. See also Buttmann, *The Shadow of the Telescope*, 189–90.

24. Spottiswoode to Gassiot, 26 May 1871, RS:MC. 9.207.

25. Gassiot to Scott, 3 June 1871, TNA:BJ 1/92.

26. RS:CM, 15 June 1871. The trust deed itself is reprinted in, for example, *The Record of the Royal Society of London for the Promotion of Natural Knowledge*, 4th ed., 134–38.

27. RS:CM, 29 June 1871.

28. BAAS:CM, 5 November 1870.

29. BAAS:CM, 2 August 1871; MMC, 3 July 1871, TNA:BJ 8/9.

30. Thomson, "Address of Sir William Thomson, Knt., LL.D., F.R.S., President," lxxxvii.

31. MacLeod, "The Support of Victorian Science," 227.

32. An example was the state-of-the-art Lick Observatory for astrophysics on Mount Hamilton, California, funded by American magnate James Lick. Osterbrock, Gustafson, and Unruh, *Eye on the Sky*.

33. Gassiot to Sharpey, 4 July 1871, RS:MS. 843.30.

34. MacLeod, "The Support of Victorian Science," 211.

35. Hall, *All Scientists Now*, 106–8.

36. Gassiot, *Remarks on the Resignation of Sir Edward Sabine K.C.B. of the Presidency of the Royal Society*.

37. White, *The Journals of Walter White* (4 January 1871): 229.

38. Barton, "'An Influential Set of Chaps,'" 67.

39. White, *The Journals of Walter White* (17 March 1871): 232.

40. Jim Burton and Malcolm Walker have both described this move by Airy in the context of the history of the Meteorological Office. Burton, "The History of the British Meteorological Office to 1905," 113–16; and Walker, *History of the Meteorological Office*, 104–6.

41. Airy to Scott, 11 December 1871, RGO 6.394.240.

42. Airy to Jeffery, 13 December 1871, RGO 6.394.244; Jeffery to Airy, 14 December 1871, RGO 6.394.245; Airy to Jeffery, 22 December 1871, RGO 6.394.247.

43. Original version with letter from Airy to Stokes, 6 January 1872, RGO 6.394.256; printed version in RGO 6.394.276.

44. Wilson, "Stokes and Kelvin, Cambridge and Glasgow, Light and Heat," 118.

45. Stokes to Airy, 8 January 1872, RGO 6.394.262; Stokes to Airy, 12 January 1872, RGO 6.394.268; RS:CM, 18 January 1872; MMC, 29 January 1872, TNA:BJ 8/9.

46. MMC, 18 December 1871, TNA:BJ 8/9.

47. MMC, 15 January 1872, TNA:BJ 8/9.

48. MMC, 29 January 1872, TNA:BJ 8/9.

49. Airy to Scott, 2 February 1872, in MMC, 5 February 1872, TNA:BJ 8/9.

50. RS:CM, 11 April and 18 April 1872.

51. Walker, *History of the Meteorological Office*, 106.

52. Higgitt, "A British National Observatory"; Chapman, "Science and the Public Good," 45–46 and 55; Alter, *The Reluctant Patron*, 80–81.

53. Airy to Jeffery, 26 December 1871, TNA:BJ 1/38; Whipple to Airy, 28 December 1871, RGO 6.394.250.

54. Airy to De La Rue, 1 January 1872, RGO 6.394.252; De La Rue to Airy, 2 January 1872, RGO 6.394.253–255.

55. Minutes of Greenwich Observatory Board of Visitors, 1 June 1872, RGO 55.

56. Airy to Stokes, 19 October 1872, RS:MC9.419; RS:CM, 31 October 1872.

57. Airy to Jeffery, 7 November 1872, TNA:BJ 1/42/402; *Report of the Astronomer Royal to the Board of Visitors*, 1873; *Report of the Astronomer Royal to the Board of Visitors*, 1874.

58. Higgitt, "A British National Observatory," 614–18.

59. Schaffer, "Where Experiments End," 276; Hutchins, *British University Observatories 1772–1939*, 284.

60. "Anniversary Meeting" [1872)], 30; RS:CM, 19 December 1872.

61. Wilfrid Airy (ed.), *Autobiography of Sir George Biddell Airy*, 303; Hall, *All Scientists Now*, 111.

62. Airy to De La Rue, 12 December 1870, RGO 6.396.17.

63. Stewart to [Whipple?], 20 December 1872, TNA:BJ 1/84.

64. Strange, "Lieutenant-Colonel Alexander Strange, FRS, examined," 76; Becker, *Unravelling Starlight*, 139–40.

65. De La Rue, "Warren De La Rue, Esq., DCL, FRS, Examined," 302.

66. Airy, "George B. Airy, Esq., CB, PRS, Examined," 93–94 and 97; Becker, *Unravelling Starlight*, 140–41.

67. Strange, "On the Insufficiency of Existing National Observatories," 240; Becker, *Unravelling Starlight*, 137.

68. Anon., "Royal Astronomical Society. Session 1871–72. Sixth Meeting, April 12th, 1872"; also cited in Becker, *Unravelling Starlight*, 138–39.

69. Hollis, "The Decade 1870–1880," 174–75; also quoted in Meadows, *Science and Controversy*, 96; also Becker, *Unravelling Starlight*, 142–43.

70. De La Rue, "Warren De La Rue, Esq., DCL, FRS, Examined," 302.

71. Becker has cited documentary evidence that Huggins and Airy corresponded in February 1872 about starting spectroscopic work at Greenwich—further suggesting that Airy was planning a cautious expansion of the work at Greenwich into astrophysics in order to forestall the plan for a separate astrophysical observatory. Becker, *Unravelling Starlight*, 141.

72. The popular science writer Stuart Clark has briefly noted that Airy's opening negotiations with De La Rue to acquire the photoheliograph set back a plan to open a new solar observatory, independent of Greenwich (Clark,

The Sun Kings, 134–35). This agrees with one strand of the argument presented here.

73. See, for example, Stewart to Whipple, 11 April 1871, TNA:BJ 1/84.

74. Katie Stewart to Whipple, 27 January 1871, TNA:BJ 1/84.

75. Clerke (also McConnell), "Whipple, George Mathews (1842–1893)"; Stewart to [Whipple?], 28 February 1871, TNA:BJ 1/84.

76. Scott to Gassiot, 22 May 1871, TNA:BJ 1/92.

77. Clerke and McConnell, "Whipple, George Mathews (1842–1893)"; BAAS:CM, 22 September 1858 and 27 June 1860; Scott to Gassiot, 22 May 1871, TNA:BJ 1/92.

78. Stewart to Sabine, 27 June 1871, TNA:BJ 1/201/8.

79. Jeffery to Whipple, 25 July 1871, TNA:BJ 1/83.

80. Savours and McConnell, "The History of the Rossbank Observatory, Tasmania," 560; Walker, *History of the Meteorological Office*, 103.

81. Stewart to Sabine, 5 July 1867, RS:Sa. 1604.

82. The baby, born on 1 August 1871, was Robert Stewart Whipple (1871–1953), later to become a well-known collector of scientific instruments and founder of the Whipple Museum of the History of Science at the University of Cambridge. Lang (also Bradley), "Whipple, Robert Stewart (1871–1953)."

83. De La Rue to Whipple, 8 July 1871, TNA:BJ 1/214/6; Gassiot to De La Rue, 7 July 1871, TNA:BJ 1/201/11.

84. Scott to Gassiot, 22 May 1871, TNA:BJ 1/92.

85. Scott to Beckley, 8 December 1871, TNA:BJ 1/214/24; KCR [1872], 40.

86. Sabine to Controller-in-Chief, War Office, 15 December 1871, TNA: BJ 3/55.

87. KCM, 3 July 1871.

88. Scott to Jeffery, 13 December 1872, TNA:BJ 1/42/441.

89. KCM, 17 April 1874; Scott to Jeffery, 7 July [18]74, TNA:BJ 1/214/132; Jeffery to Scott, 11 November 1875, TNA:BJ 1/202/231; Scott to Jeffery, 27 November 1875, TNA:BJ 1/214/252.

90. KCM, 25 July 1873.

91. KCM, 18 December 1874; Sabine to Bergsma, 22 November 1873, RS:Sa. 1250; Scott to [Charles] Meldrum, 13 January 1876, TNA:BJ 1/214/258. See also Good, "Sir Edward Sabine."

92. KCM, 19 November 1875.

93. Scott to Meldrum, 13 January 1876, TNA:BJ 1/214/258.

94. KCM, 19 November 1875, 25 February 1876, and 14 November 1876.

95. Scott to E. Quetelet, 20 January [18]74, TNA:BJ 1/214/95.

96. KCM, 23 December 1879; Gooday, "Precision Measurement and the Genesis of Physics Teaching Laboratories in Victorian Britain" (1989), chap. 5, 26–37.

97. KCR [1876], 373.

98. KCR [1882], 352; Morus, *When Physics Became King*, 241; Schaffer, "Late Victorian Metrology and Its Instrumentation," 36.

99. MMC, 20 November 1876, TNA:BJ 8/9.

100. Anderson, *Predicting the Weather*, 144; Walker, *History of the Meteorological Office*, 113–14.

101. MMC, 9 July 1877, TNA:BJ 8/9.

102. MMC, 24 January 1879, TNA:BJ 8/10; KCR [1879], 445.

103. KCM, 28 April 1876; KCR [1876], 373; KCR [1879], 447.

104. KCR [1881], 84; KCR [1882], 348.

105. KCR [1884], 466–67; KCR [1885], 318.

106. KCM, 3 November and 1 December 1871.

107. Roscoe to Scott, 16 July 1875, TNA:BJ 1/202/206.

108. Scott, "The History of the Kew Observatory," 76; KCR [1872], 43.

109. Galton, *Memories of My Life*, 227; Galton, "Description of the Process of Verifying Thermometers at the Kew Observatory"; KCR [1875], 106.

110. McConnell, *King of the Clinicals*, 22; Lawrence, "Incommunicable Knowledge," esp. 515.

111. Galton, *Memories of My Life*, 227.

112. KCM, 16 July 1874.

113. KCR [1875], 110 and 106.

114. Scott to C. R. Gore, Office of Woods and Forests, 6 March 1875, TNA:BJ 1/214/178.

115. Printed notice, dated 29 January 1876, pasted into KCM, 28 January 1876.

116. KCM, 31 July 1876; Francis Galton to Scott, 24 April 1877, TNA:BJ 1/202/320; KCM, 27 July 1877.

117. KCM, 29 November 1878 and 19 December 1878.

118. KCM, 29 November and 19 December 1878. KCR [1879], 453, refers to her as "a special assistant, H. Clements," without mentioning that she was female.

119. A. D. Morrison-Low, "Women in the Nineteenth-Century Scientific Instrument Trade," esp. 101 and 103.

120. KCM, 24 November 1880; KCM, 2 November 1881.

121. KCM, 30 March 1883.

122. McConnell, *King of the Clinicals*, 23.

123. KCM, 23 March and 28 May 1880.

124. Barrell, "Kew Observatory and the National Physical Laboratory," 175; KCR [1880], 124.

125. KCM, 26 May 1882.

126. KCM, 29 April 1875; KCR [1875], 107; F. J. Britten (British Horological Institute) to Scott, 2 June 1875, TNA: BJ 1/201/197.

127. "Ebor.," "Watch Rate Papers," 56; "Preliminary Report on the Proposal to Establish a System of Rating of Superior Watches at the Kew Observatory," signed by Whipple, dated 29 March 1882, TNA:BJ 1/206/716; KCM, 4 February 1881, 24 November 1881, and 23 December 1881.

128. KCM, 27 January 1882.

129. KCM, 26 January 1883 and 23 February 1883; "Scheme for the Rating of Watches at the Kew Observatory, Richmond," signed by Whipple, dated 27 February 1882 (should read 1883 in context of KCM), TNA:BJ 1/206/775; KCR [1883], 93–94; KCR [1885], 320.

130. KCR [1885], 324.

131. KCR [1881], 87.

132. RS:CM, 18 January 1883 and 15 February 1883.

133. [Spottiswoode] to Stewart (draft letter, no date, probably May 1872), TNA:BJ 1/214/37.

134. Maunder, *The Royal Observatory Greenwich*, 232–33.

135. Davis, "Weather Forecasting and the Development of Meteorological Theory at the Paris Observatory, 1853–1878," 381.

136. MacLeod, "Introduction," xii; MacLeod, "Whigs and Savants," esp. 77–81.

CHAPTER 5: KEW OBSERVATORY AND THE ORIGINS OF THE NATIONAL PHYSICAL LABORATORY, 1885–1900

Epigraphs: Lodge to BAAS Committee on a National Physical Laboratory, 22 February 1893, TNA:BJ 1/210/1151F; "On the Establishment of a National Physical Laboratory," 84.

1. Moseley, "Science, Government & Industrial Research," 39–66; Moseley, "The Origins and Early Years of the National Physical Laboratory"; Alter, *The Reluctant Patron*, 138–49; Pyatt, *The National Physical Laboratory*, 12–33; Magnello, *A Century of Measurement*, 11–30. On the origins and development of the PTR, see Cahan, *An Institute for an Empire*, 10–28.

2. KCR [1885], 325; KCR [1899], 352; KCR [1890], 501.

3. RS:CM, 12 March 1891.

4. KCM, 27 February 1891; RS:CM, 21 January 1892.

5. KCR [1893], 307.

6. KCR [1895], 392; KCR [1896], 106.

7. Moseley, "The Origins and Early Years of the National Physical Laboratory," 234 and 249.

8. KCM, 25 May 1883; KCR [1892], 322. A certificate accompanying a sextant tested at Kew, now in the collection of the Museum of the History of Science, University of Oxford (Inv. No. 45880) is dated December 1892 and bears Whipple's signature.

9. RS:CM, 19 January 1893.

10. Francis Galton to Bank of England, 8 February 1893, TNA:BJ 1/210; KCM, 24 February 1893. The advertisement appears on the front page of *The Athenaeum*, no. 3412 (18 March 1893).

11. "Information to Applicants for the Post of Superintendent," dated 1 March 1893, inserted in KCM, 24 February 1893.

12. KCM, 24 March and 28 April 1893.

13. H. J., "Herbert Tomlinson—1845–1931"; Taylor and Havelock, "William Cecil Dampier 1867–1952"; S[impson], "Charles Chree, 1860–1928"; "Mr. T. H. Blakesley."

14. KCM, 28 April 1893.

15. Ranked list of applicants for the Kew superintendent's position (undated, but likely to have been drawn up on or around 21 April 1893, the date of the selection subcommittee's meeting), TNA:BJ 1/210/1154.

16. S[impson], "Charles Chree, 1860–1928," viii.

17. Sviedrys, "The Rise of Physical Science at Victorian Cambridge," 143.

18. Kinder, "Edward Walter Maunder FRAS (1851–1928)," 23–24.

19. RS:CM, 16 February 1893.

20. Sviedrys, "The Rise of Physics Laboratories in Britain," esp. 422–27; Gooday, "Precision Measurement and the Genesis of Physics Teaching Laboratories in Victorian Britain" (1990).

21. Grier, *When Computers Were Human*, 53.

22. "Information to Applicants for the Post of Superintendent," dated 1 March 1893, inserted in KCM, 24 February 1893.

23. Lodge to BAAS Committee on National Physical Laboratory, 22 February 1893, TNA:BJ 1/210/1151F.

24. Glazebrook to Galton, 23 February [1893], TNA:BJ 1/210/1151F.

25. Schaffer, "Late Victorian Metrology and Its Instrumentation."

26. KCM, 17 December 1897; KCM, 15 December 1899.

27. KCM, 24 February 1893.

28. E. G. Constable to Secretary, Kew Committee, 21 January 1893, TNA:BJ 1/209/1150.

29. E. G. Constable, W. Hugo, J. Foster, J. Gunter, and W. Boxall to chairman and members of Kew Observatory Committee, 14 February 1898, TNA:BJ 1/213.

30. "Information to Applicants for the Post of Superintendent," dated 1 March 1893, inserted in KCM, 24 February 1893.

31. Gough to Whipple, 19 May 1890, TNA:BJ 1/209.

32. KCM, 15 December 1899.

33. Document for inspection of Kew Observatory by General Board of NPL, October 1899, TNA:BJ 1/213/1361.

34. Grier, *When Computers Were Human*, 50–52.

35. Hutchins, *British University Observatories 1772–1939*, 69.

36. Chree, "Description of the Kew Observatory."

37. Document for inspection of Kew Observatory by General Board of NPL, October 1899, TNA:BJ 1/213/1361. It is not clear whether the staff wore ribbons as a matter of routine or just for the inspection, but even in the latter case it demonstrates the hierarchical staff structure.

38. Scrase, "Some Reminiscences of Kew Observatory in the Twenties," 181.

39. Chapman, "Sir George Airy (1801–1892) and the Concept of International Standards in Science, Timekeeping and Navigation," 321. See also Smith, "A National Observatory Transformed," 13–17, and Grier, *When Computers Were Human*, 52–53.

40. KCR [1890], 495.

41. KCR [1887], 212; KCR [1889], 475.

42. S[impson], "Charles Chree, 1860–1928," ix–xi.

43. KCR [1891], 167.

44. Higgitt, "A British National Observatory," esp. 625.

45. S[impson], "Charles Chree, 1860–1928," xii.

46. Milne to Chree, 14 June 1896, TNA:BJ 1/212/1256.

47. Morus, *When Physics Became King*, 236; Gooday and Low, "Technology Transfer and Cultural Exchange," esp. 121–27 on John Milne.

48. Herbert-Gustar and Nott, *John Milne*, 122–33.

49. KCM, 23 October 1896; Milne to Chree, 14 December 1896, TNA:BJ 1/212/1256; M. Foster (Secretary RS) to Chree, 3 June 1897, TNA:BJ 1/212/1285; Chree to John Perry, 23 June 1897, TNA:BJ 1/212; Williams, *The Precision Makers*, 22; KCR [1898], 5–6.

50. See, for example, KCR [1894], 503.

51. KCR [1890], 493–94; Walker, *History of the Meteorological Office*, 120–21.

52. KCR [1888], 78.

53. Anderson, *Predicting the Weather*, 225; KCM, 30 January 1891; KCR [1891], 156–57.

54. KCR [1894], 507; Tucker, "Photography as Witness, Detective, and Impostor," 381–89.

55. KCR [1894], 503; KCR [1897], 164.

56. BAAS:CM, 16 November 1860.

57. KCR [1881], 83.

58. KCR [1894], 503.

59. Chree, "Observations on Atmospheric Electricity at the Kew Observatory," 132.

60. KCR [1899], 344; Longair, "Charles Thomson Rees Wilson."

61. KCR [1886]; KCR [1899].

62. KCR [1889].

63. KCR [1881]; KCR [1899].

64. KCR [1888], 83; KCR [1889], 480; KCM, 29 November 1889.

65. Marder, *The Anatomy of British Sea Power*, 119–43; Fanning, *Steady as She Goes*, 133–34.

66. E. [Risham?] (53 Parliament Street, London SW) to Whipple, 11 November 1886, TNA:BJ 1/208/972; KCR [1887], 218.

67. KCR [1888], 83.

68. Ettrick Creak to Whipple, 23 May 1887, TNA:BJ 1/208/996. For Creak's career at the Admiralty Compass Department, see Fanning, *Steady as She Goes*, esp. 128–35 and 152–54.

69. KCM, 6 November and 27 November 1885.

70. KCR [1886], 409; specimen of a Kew chronometer rating certificate, accompanying letter from George Whipple to William Christie, 16 February 1888, RGO 7.95.

71. KCR [1886], 409; KCR [1898].

72. Whipple to Christie, 16 February 1888, RGO 7/95; Christie to Whipple, 21 February 1888, RGO 7/95.

73. KCM, 30 January, 26 March, and 24 April 1891; KCR [1891], 160.

74. Lewis to Tizard, 18 May 1893, RGO 7.104.683; Tizard to Lewis, 20 May 1893, RGO 7.104.684; Lewis to Tizard, 30 May 1893, RGO 7.104.685.

75. KCR [1891], 160.

76. KCM, 28 April 1893; KCR [1893], 314; KCR [1894], 506.

77. KCR [1886], 408.

78. Bonniksen to Chree, 3 December 1897, TNA:BJ 1/212; also copy of

an article from December 1897 *Horological Journal* signed by S. Smith & Son, TNA:BJ 1/212. The head of Bonniksen's letter cited here advertised that sixty of his watches were marked "especially good" at Kew in 1896.

79. KCM, 15 December 1899 and 19 January 1900.

80. KCM, 25 January 1895.

81. Middleton, *A History of the Thermometer and Its Use in Meteorology*, 179–80; Cattermole and Wolfe, *Horace Darwin's Shop*, esp. 185–88; J. A. Harker to Chree, 11 May 1896, TNA: BJ 1/212/1251.

82. KCR [1895], 387–88.

83. Harker to Chree, 11 May 1896, TNA: BJ 1/212/1251; KCR [1896], 100; KCR [1897], 165–66.

84. Crease, *World in the Balance*, 136–38.

85. KCR [1899], 345; KCM, 10 November 1899.

86. KCM, 29 March 1895; Cattermole and Wolfe, *Horace Darwin's Shop*, 49.

87. KCR [1897], 166.

88. KCM, 21 June 1895; Glazebrook (also McConnell), "Darwin, Horace"; Cattermole and Wolfe, *Horace Darwin's Shop*, 47.

89. KCM, 21 October 1898; KCR [1898], 7; Pyatt, *The National Physical Laboratory*, 32 and 230.

90. Chree, "Description of the Kew Observatory."

91. Chree, "Description of the Kew Observatory"; KCM, 25 March 1887; KCR [1888], 84; Lord Rayleigh to Francis Galton, 9 May 1891, TNA:BJ 1/209/1114; KCR [1892], 331; Document for inspection of Kew Observatory by General Board of NPL, October 1899, TNA:BJ 1/213/1361.

92. See, for example, Moseley, "Science, Government & Industrial Research," 22; Alter, *The Reluctant Patron*, 138–49; Magnello, *A Century of Measurement*, 12–14.

93. Thomson, "Address of Sir William Thomson, Knt., LL.D., F.R.S., President," lxxxviii.

94. Lodge, "Section A.—Mathematical and Physical Science," esp. 549–51.

95. The committee was formally described as "the Committee on a National Physical Laboratory" (*BAAS Report* [1892], xiii.). This seems to be the earliest use of the exact term "National Physical Laboratory."

96. Moseley, "Science, Government & Industrial Research," 44; Moseley, "The Origins and Early Years of the National Physical Laboratory," 224–25; Alter, *The Reluctant Patron*, 139.

97. Lodge to Francis Galton, 21 February 1893, TNA:BJ 1/210/1151F.

98. Lodge to BAAS Committee on National Physical Laboratory, 22 February 1893, TNA:BJ 1/210/1151F.

99. Lodge to Galton, 22 February 1893, TNA:BJ 1/210/1151F.

100. Schuster to Galton, 22 February 1893, TNA:BJ 1/210/1151F; Thomson to Galton, 23 February 1893, TNA:BJ 1/210/1151F; Glazebrook to Galton, 23 February [1893], TNA:BJ 1/210/1151F.

101. KCM, 24 February 1893.

102. Galton, Douglas, "Address by Sir Douglas Galton, KCB, DCL, FRS, President," 34.

103. Galton, Douglas, "Address by Sir Douglas Galton, KCB, DCL, FRS, President," 32.

104. Galton, Douglas, "On the Reichsanstalt, Charlottenburg, Berlin," 607–08.

105. Anon., "On the Establishment of a National Physical Laboratory," 84.

106. Document headed "NATIONAL PHYSICAL LABORATORY," enclosed with Douglas Galton to Secretaries of the Royal Society, 21 October 1896, RS:MC. 16.335.

107. Douglas Galton to Secretaries of the Royal Society, 21 October 1896, RS:MC. 16.335.

108. Douglas Galton to Michael Foster, 21 October 1896, RS:MC. 16.334.

109. KCM, 23 October 1896; Francis Galton to Michael Foster, 24 October 1896, RS:MC. 16.337.

110. RS:CM, 5 November 1896.

111. *Report of the Committee Appointed by the Treasury to Consider the Desirability of Establishing a National Physical Laboratory*, iii.

112. Moseley, "Science, Government & Industrial Research," 58.

113. *Minutes of Evidence of Taken before the Committee Appointed by the Treasury to Consider the Desirability of Establishing a National Physical Laboratory*, iii–iv.

114. *Report of the Committee Appointed by the Treasury to Consider the Desirability of Establishing a National Physical Laboratory*, 6.

115. E. W. Hamilton to [Lord Rayleigh], 7 October 1898, TNA:BJ 1/213/1327.

116. M. Foster and A. W. Rücker (Secretaries of Royal Society), to Secretary to HM Treasury, 28 November 1898, RS:MS.538. The Treasury soon agreed to continue the £400 Meteorological Office grant (Francis Mowatt, Treasury, to President of the Royal Society, 7 December 1898, in RS:CM, 8 December 1898).

117. Noble to Rücker, 19 October 1898, RS:MS.538. Noble was a member of the government committee that reported on the NPL in July 1898.

118. RS:CM, 3 November 1898.

119. "NATIONAL PHYSICAL LABORATORY. Draft Scheme of Organization. Revised at a Meeting of the Committee on January 11th. 1899," TNA:BJ 1/213.

120. KCM, 10 November and 1 December 1899. On Kempe, see Geikie, "Sir Alfred Bray Kempe, 1849–1922," v–ix.

121. *NPL Report* [1901].

122. Royal Society: NPL:ECM, 5 May 1899; RS:CM, 15 June 1899.

123. Royal Society: NPL:ECM, 5 July 1899; Boyle to Rücker, 7 June 1899, RS:MS.538.

124. Schaffer, "Late Victorian Metrology and Its Instrumentation," 41; Moseley, "Glazebrook, Sir Richard Tetley (1854–1935)."

125. Moseley, "Glazebrook, Sir Richard Tetley (1854–1935)."

126. KCM, 10 November 1899; RS:CM, 15 February 1900.

127. *Hansard*, HC Deb 07 May 1900 vol. 82 cc882–3; *Hansard*, HC Deb 14 May 1900 vol. 83 cc39–40; *Hansard*, HC Deb 28 May 1900 vol. 83 c1511; Moseley, "Science, Government & Industrial Research," 75.

128. Francis Mowatt (Treasury) to President of the Royal Society, 30 October 1900, RS:MS.538.

129. RS:CM, 25 October 1900; Pyatt, *The National Physical Laboratory*, 12; Lord Esher (HM Office of Works) to Secretary, Royal Society, 22 December 1900, RS:MS.538.

130. *NPL Report*, 1901.

131. Galton, Francis, *Memories of My Life*, 228.

132. Glazebrook, *Early Days at The National Physical Laboratory*, 3.

133. *BAAS Report* [1896], 84.

CHAPTER 6: "AN EPOCH IN THE HISTORY OF KEW": THE END OF THE VICTORIAN KEW OBSERVATORY, 1900–1910

Epigraph: Shaw to Glazebrook, 7 May 1909, in NPL:ECM, 21 May 1909.

1. Edgerton, *Warfare State*, esp. 1–2.

2. NPL:ECM, 28 February 1900.

3. Pearson, *The Life, Letters and Labours of Francis Galton, Vol. 2*, 60.

4. *NPL Report* [1902], 34; *NPL Report* [1909], 23–24.

5. Chree to Harrison (Assistant Secretary, Royal Society), 23 January 1902, TNA:BJ 1/128.

6. Chree to Glazebrook, 19 July 1909, TNA:BJ 1/139.

7. Chree to Carey Foster, 29 November 1899, TNA:BJ 1/122.

8. NPL:ECM, 24 April 1903.

9. *NPL Report* [1909], 23; *NPL Report* [1906], 10. Dates when individuals first started work at Kew can be found directly from the staff lists that appear in the Kew Committee Reports from 1872.

10. *NPL Report* [1902], 34; *NPL Report* [1909], 24.

11. NPL:ECM, 13 March 1901; MMC, 14 July 1909.

12. NPL:ECM, 19 February 1904 and 13 March 1901.

13. Moseley, "Science, Government & Industrial Research," 83–85; Moseley, "The Origins and Early Years of the National Physical Laboratory," 237–38 and 245.

14. Chree to *Richmond & Twickenham Times* and Chree to *The Surrey Comet*, 6 February 1902, TNA:BJ 1/128; staff lists in *NPL Reports*, 1901–1910.

15. *NPL Report* [1906], 10.

16. Williams, *The Precision Makers*, 48.

17. Memorandum (dated 2 February 1910) in MMC, 2 March 1910.

18. MMC, 6 November 1907.

19. *NPL Report* [1908], 92. For background to the international cooperation in meteorology, see Walker, *History of the Meteorological Office*, 109–12.

20. *NPL Report* [1901], 21.

21. *NPL Report* [1906], 57; Mason, "Simpson, Sir George Clarke (1878–1965)."

22. Walker, *History of the Meteorological Office*, 124–29; Burton, "Shaw, Sir (William) Napier (1854–1945)."

23. Quoted in Anderson, *Predicting the Weather*, 12.

24. Walker, *History of the Meteorological Office*, 151–56.

25. "The National Physical Laboratory. Report on the Observatory Department for the Year Ending December 31, 1900," 425; *NPL Report* [1910], 94; *NPL Report* [1909], 94.

26. *NPL Report* [1912], 116.

27. *NPL Report* [1905], 52.

28. *NPL Report* [1901], 5.

29. *NPL Report* [1909], 95; *NPL Report* [1911], 93, *NPL Report* [1912], 116.

30. *NPL Report* [1902], 30.

31. *NPL Report* [1903], 4–5.

32. MMC, 3 November 1909.

33. *NPL Report* [1901], 25; *NPL Report* [1906], 57.

34. KCM, 15 December 1899.

35. NPL:ECM, 16 December 1904; RS:CM, 19 January 1905 and 16 March 1905.

36. NPL:ECM, 15 February 1907; RS:CM, 30 April 1908; RS:CM, 30 May 1907.

37. *BAAS Report* [1860], xli.

38. KCM, 7 February 1898.

39. KCR [1898], 12.

40. *NPL Report* [1901], 18–19.

41. NPL:ECM, 11 December 1901.

42. *NPL Report* [1902], 25–26. These tables were similarly not included in subsequent reports.

43. KCM, 20 October 1899.

44. NPL:ECM, 18 December 1903.

45. NPL:ECM, 17 April 1901.

46. NPL:ECM, 11 December 1901.

47. *NPL Report* [1903], 29; NPL:ECM, 18 December 1903.

48. RS:CM, 10 December 1903; RS:CM, 21 January 1904.

49. NPL:ECM, 16 November 1906.

50. *NPL Report* [1907], 78 and 10.

51. *NPL Report* [1908], 95–96 and 6–7; NPL:ECM, 23 May 1910.

52. NPL:ECM, 16 November 1906.

53. C. V. B., [Chree] and (Turner), "G. W. Walker, 1874–1921."

54. MMC, 5 November 1902 and 19 November 1902, TNA:BJ 8/15; RS:CM, 22 January 1903.

55. *NPL Report* [1908], 96–97.

56. *NPL Report* [1908], 92–93.

57. MMC, 6 November 1907.

58. Glazebrook to Shaw, 30 January 1908, in MMC, 5 February 1908.

59. MMC, 5 February 1908.

60. Magnello, *A Century of Measurement*, 41; also Moseley, "Science, Government & Industrial Research," 90–91; Moseley, "The Origins and Early Years of the National Physical Laboratory," 241.

61. Glazebrook to Shaw, 12 February 1908, in MMC, 4 March 1908; Shaw to Glazebrook, 19 February 1908, in MMC, 4 March 1908.

62. Shaw to Glazebrook, 7 May 1909, in NPL:ECM, 21 May 1909. (Letter is also in MMC, 5 May 1909.)

63. NPL:ECM, 21 May 1909.

64. RS:CM, 17 June 1909; MMC, 14 July 1909.

65. MMC, 14 July 1909.

66. Kempe to Rücker, 21 October 1898, RS:MS.538.

67. MMC, 14 July 1909.

68. "Kew Observatory. Draft Report of a Joint Committee of Representatives of the Council of the Royal Society and of the Executive Committee of the National Physical Laboratory," in MMC, 3 November 1909.

69. NPL:ECM, 15 October 1909.

70. RS:CM, 28 October 1909; MMC, 3 November 1909; RS:CM, 9 December 1909.

71. MMC, 1 December 1909.

72. G. H. Murray (Treasury) to Secretaries, RS, 10 January 1910, in RS:CM, 20 January 1910.

73. "The National Physical Laboratory. Report on the need for additional buildings, specially in regard to the letter from the Director of the Meteorological Office, dated May 7, 1909, as to the relations between Kew Observatory and the Office. Adopted by the Executive Committee, December 17, 1909," RS:MS.538, esp. 6.

74. RS:CM, 20 January 1910.

75. MMC, 2 February 1910.

76. RS:CM, 17 February 1910.

77. Shaw to Treasury, 21 February 1910, in MMC, 2 March 1910.

78. RS:CM, 17 February 1910; Shaw to Treasury, 21 February 1910, in MMC, 2 March 1910.

79. G. H. Murray (Treasury) to Shaw, 26 February 1910, in MMC, 2 March 1910; Murray (Treasury) to Secretaries, RS, 26 February 1910, in RS:CM, 17 March 1910.

80. NPL:ECM, 23 May 1910.

81. RS:CM, 17 March 1910; RS:CM, 28 April 1910.

82. NPL:ECM, 17 June 1910.

83. "Kew and Eskdale Muir Observatories and the Meteorological Office."

84. Glazebrook to Larmor, 21 July 1910, RS:MS.538.

85. NPL:ECM, 16 December 1910.

86. Glazebrook to Larmor, 21 July 1910, RS:MS.538. On Haldane and aeronautics research at the NPL, see Moseley, "Science, Government & Industrial Research," 107-9.

87. "Report of the Council" (1912), 186; Murray (Treasury) to Secretary, Royal Society, 17 June 1911, in NPL:ECM, 21 July 1911; Magnello, A Century of Measurement, 51-52; Moseley, "Science, Government & Industrial Research," 111-14; Moseley, "The Origins and Early Years of the National Physical Laboratory," 245-47.

88. *NPL Report* [1912], 63 and 118; *NPL Report* [1913–1914], 7.

89. G. H. Murray (Treasury) to Secretaries, RS, 10 January 1910, in RS:CM, 20 January 1910; NPL:ECM, 17 June 1910.

90. Schaffer, "Late Victorian metrology and its instrumentation," 37; Kim, "J. J. Thomson and the emergence of the Cavendish School, 1885–1990," 204–5.

91. *NPL Report* [1912], 62–63; Pearson, *The Life, Letters and Labours of Francis Galton, Vol. 2*, 59; Chree to Glazebrook, 18 April 1912, TNA: BJ 1/142.

92. See, for example, the certificate for barometer M.O. 1154, dated 17 September 1915, in "Photographs relating to the history of Kew Observatory, the museums at Kew and at the Meteorological Office," National Meteorological Archive, Exeter, UK, shelf mark: ARCHIVE Z28.L4.

93. Magnello, *A Century of Measurement*, 122.

94. NPL:ECM, 17 June 1910.

95. Shaw to Director of NPL, 31 July 1914, in MMC, 28 October 1914; Reid, "*We're certainly not afraid of Zeiss,*" 91 n. 147; Mörzer-Bruyns, *Sextants at Greenwich*, 53.

96. *NPL Report* [1912], 63; *NPL Report* [1913], 96–97.

97. *NPL Report* [1912], 116; NPL:ECM, 21 June 1912.

98. NPL:ECM, 19 March 1918; Kew staff photograph, May 1927, in "Plaques and portraits," National Meteorological Archive, Exeter, UK, shelf mark: ARCHIVE Z33.E3.

99. Chree to [Robert H. Scott?], 13 March 1893, in Kew Box File 2—"Golden Oldies," National Meteorological Archive, Exeter, UK.

100. Walker, *History of the Meteorological Office*, 279–82.

101. Meteorological Office, *The Observatories' Year Book 1922*, 271; Meteorological Office, *The Observatories' Year Book 1967*, iv; International Seismological Centre website, http://www.isc.ac.uk/, accessed 22 September 2016.

102. Harrison, "The British radiosonde"; Scrase, "Some reminiscences of Kew Observatory in the Twenties," 183; Blackwell, "The Bicentenary of Kew Observatory," 236–38.

103. Scrase, "Some reminiscences of Kew Observatory in the Twenties," 180.

104. Meteorological Office, *The Observatories' Year Book 1922*, 258.

105. "Photographs relating to the history of Kew Observatory, the museums at Kew and at the Meteorological Office," National Meteorological Archive, Exeter, UK, shelf mark: ARCHIVE Z28.L4.

106. *Meteorological Magazine* 98, no. 1163 (June 1969).

107. Walker, *History of the Meteorological Office*, 397–98.

108. [Mason], "Foreword By the Director-General of the Meteorological Office," 161; Blackwell, "The Bicentenary of Kew Observatory," 238; Walker, *History of the Meteorological Office*, 398.

109. Walker, *History of the Meteorological Office*, 398–400; Galvin, "Kew Observatory," 483; "Cruel Cuts at the Meteorological Office," 137; Mayes, "Kew Observatory, 1769 to 1980."

110. *Year-Book of the Royal Society 1912*, 183.

111. Martin, "The Gassiot Committee of the Royal Society and Meteorological Research"; Massey and Robins, *History of British space science*, esp. 4–8; Stewart, "Development of rocket and satellite experiments at Kew Observatory, 1959–61."

112. Martin, "The Gassiot Committee of the Royal Society and Meteorological Research," 212–13; *Year Book of the Royal Society 1969*, 297; *Year Book of the Royal Society 1981*, 314.

113. Shaw to Glazebrook, 7 May 1909, in NPL:ECM, 21 May 1909.

114. Moseley, "Science, Government & Industrial Research," 85.

115. Peden, "Murray, Sir George Herbert (1849–1936)"; Headlam and Thomas (also Booth), "Heath, Sir Thomas Little (1861–1940)"; Peden, *The Treasury and British Public Policy*, 72.

CONCLUSION

Epigraphs: Whipple, "Some Aspects of the Early History of Kew Observatory," 135; Hufbauer, *Exploring the Sun*, 49.

1. Hughes, "Redefining the Context," esp. 299–300, has pointed to the richness of physics teaching in the London colleges and elsewhere at the beginning of the twentieth century, in contrast with the received view of the primacy of the Cavendish Laboratory.

2. Hughes, "Redefining the Context," esp. 271–72.

3. Moseley, "The Origins and Early Years of the National Physical Laboratory," 237–38.

4. Smith, "A National Observatory Transformed," 18.

5. "On the Establishment of a National Physical Laboratory," 84.

6. Moseley, "The Origins and Early Years of the National Physical Laboratory," esp. 249.

7. *Royal Commission on Scientific Instruction and the Advancement of Science. Eighth Report*, 49–50; also cited in Anderson, *Predicting the Weather*, 141.

8. Hull, "War of Words," esp. 480.

9. Moseley, "Science, Government & Industrial Research," 39–66 and 359; Moseley, "The Origins and Early Years of the National Physical Laboratory."

BIBLIOGRAPHY

ARCHIVAL SOURCES

Bodleian Library, Oxford, UK

BAAS PAPERS: BAAS MINUTES OF COUNCIL, BAAS:CM

Harry Ransom Center, University of Texas at Austin, USA

CORRESPONDENCE OF SIR JOHN HERSCHEL: TXU

Institute of Engineering and Technology, London, UK

PAPERS OF SIR FRANCIS RONALDS, JOURNAL OF COPY LETTERS AND DIARY
ENTRIES, IET
S.C.MSS.1/4/17B

Meteorological Office, National Meteorological Archive, Exeter, UK

KEW COMMITTEE MINUTES, 1849–1900
"KEW DIARY," 27 AUGUST 1850–31 OCTOBER 1851

The National Archives, Kew, UK

MINUTES OF THE METEOROLOGICAL COMMITTEE (1867–1877), METEORO-
LOGICAL COUNCIL (1877–1905), AND METEOROLOGICAL COMMITTEE
(1905–1910), TNA:BJ 8
PAPERS OF KEW OBSERVATORY, TNA:BJ 1
PAPERS OF SIR EDWARD SABINE, TNA:BJ 3

Royal Greenwich Observatory Archives, Cambridge University Library,
Cambridge

PAPERS OF GEORGE AIRY, RGO 6
PAPERS OF WILLIAM CHRISTIE, RGO 7

Royal Society, London, UK

CORRESPONDENCE OF SIR EDWARD SABINE, RS:SA
MINUTES OF COMMITTEE OF PHYSICS AND METEOROLOGY, 1839–1845,
RS:CMB.284
MISCELLANEOUS CORRESPONDENCE, RS:MC
MISCELLANEOUS MANUSCRIPTS, RS:MM
NATIONAL PHYSICAL LABORATORY PAPERS, RS:MS.538

NATIONAL PHYSICAL LABORATORY, MINUTES OF EXECUTIVE COMMITTEE

PAPERS OF SIR JOHN HERSCHEL, RS:HS

ROYAL SOCIETY, MINUTES OF COUNCIL, RS:CM

SECRETARIAL CORRESPONDENCE, 1854–1871, RS:MS.843

PUBLISHED SOURCES

Airy, G. B. "Account of Observations of the Total Solar Eclipse of 1860, July 18, made at Hereña, near Miranda de Ebro; with a notice of the general proceedings of 'The Himalaya expedition for Observation of the Total Solar Eclipse.'" *Monthly Notices of the Royal Astronomical Society* 21 (1860): 1–16.

Airy, G. B. "George B. Airy, Esq., CB, PRS, examined." In *Royal Commission on Scientific Instruction and the Advancement of Science. Eighth Report*, 93–100. 1875.

Airy, G. B. *Report of the Astronomer Royal to the Board of Visitors, Read at the Annual Visitation of the Royal Observatory, Greenwich, June 5, 1841*. London: Royal Observatory, Greenwich, 1841.

Airy, G. B. *Report of the Astronomer Royal to the Board of Visitors, Read at the Annual Visitation of the Royal Observatory, Greenwich, 1847, June 5*. London: Royal Observatory, Greenwich, 1847.

Airy, G. B. *Report of the Astronomer Royal to the Board of Visitors, Read at the Annual Visitation of the Royal Observatory, Greenwich, 1848, June 3*. London: Royal Observatory, Greenwich, 1848.

Airy, G. B. *Report of the Astronomer Royal to the Board of Visitors, Read at the Annual Visitation of the Royal Observatory, Greenwich, 1851, June 7*. London: Royal Observatory, Greenwich, 1851.

Airy, G. B. *Report of the Astronomer Royal to the Board of Visitors of the Royal Observatory, Greenwich, Read at the Annual Visitation of the Royal Observatory, 1873, June 7*. London: Royal Observatory, Greenwich, 1873.

Airy, G. B. *Report of the Astronomer Royal to the Board of Visitors of the Royal Observatory, Greenwich, Read at the Annual Visitation of the Royal Observatory, 1874, June 6*. London: Royal Observatory, Greenwich, 1874.

Airy, Wilfrid, ed. *Autobiography of Sir George Biddell Airy, K.C.B., M.A., LL.D., D.C.L., F.R.S., F.R.A.S., Honorary Fellow of Trinity College, Cambridge, Astronomer Royal from 1836 to 1881*. Cambridge: Cambridge University Press, 1896.

Alter, Peter. *The Reluctant Patron: Science and the State in Britain 1850–1920*. Oxford: Berg, 1987.

Anderson, Katharine. *Predicting the Weather: Victorians and the Science of Meteorology*. Chicago: University of Chicago Press, 2005.

"Anniversary Meeting." *Proceedings of the Royal Society* 21 (1872): 21–36.

Ashworth, William J. "John Herschel, George Airy and the Roaming Eye of the State." *History of Science* 36 (1998): 151–78.

Astronomical and Magnetical and Meteorological Observations Made at the Royal Observatory, Greenwich, in the Year 1839. London: J. Murray, 1840.

Aubin, David. "A history of observatory sciences and techniques." In *Astronomy at the Frontiers of Science*, edited by Jean-Pierre Lasota, 109–21. Heidelberg: Springer, 2011.

Aubin, David, Charlotte Bigg, and H. Otto Sibum, eds. *The Heavens on Earth: Observatories and Astronomy in Nineteenth-Century Science and Culture*. Durham, NC: Duke University Press, 2010.

Aubin, David, Charlotte Bigg, and H. Otto Sibum. "Introduction: Observatory Techniques in Nineteenth-Century Science and Society." In *The Heavens on Earth: Observatories and Astronomy in Nineteenth-Century Science and Culture*, edited by David Aubin, Charlotte Bigg, and H. Otto Sibum, 1–32. Durham, NC: Duke University Press, 2010.

Babbage, Charles. *Reflections on the Decline of Science in England, and on Some of its Causes*. London: B. Fellowes and J. Booth, 1830.

Baguley, David. *Napoleon III and His Regime: An Extravaganza*. Baton Rouge: Louisiana State University Press, 2000.

Barrell, H. "Kew Observatory and the National Physical Laboratory." *Meteorological Magazine* 98 (1969): 171–80.

Bartholomew, C. F. "The Discovery of the Solar Granulation." *Quarterly Journal of the Royal Astronomical Society* 17 (1976): 263–89.

Barton, Ruth. "'An Influential Set of Chaps': The X Club and Royal Society Politics 1864–85." *British Journal for the History of Science* 23 (1990): 53–81.

Becker, Barbara J. *Unravelling Starlight: William and Margaret Huggins and the Rise of the New Astronomy*. Cambridge: Cambridge University Press, 2011.

Beckley, R. "Description of a Self-recording Anemometer." In *Report of the Twenty-Eighth Meeting of the British Association for the Advancement of Science; Held at Leeds in September 1858*, 306–7. London: John Murray, 1859.

Bennett, J. A. *Church, State and Astronomy in Ireland: 200 Years of Armagh Observatory*. Belfast: The Armagh Observatory, 1990.

Bennett, J. A. "George Biddell Airy and Horology." *Annals of Science* 37 (1980): 269–85.

Blackwell, M. J. "The Bicentenary of Kew Observatory" *Weather* 24 (1969), pp: 235–38.

Board of Trade. *Report of a Committee Appointed to Consider Certain Questions Relating to the Meteorological Department of the Board of Trade*, 1866.

"British Association for the Advancement of Science." *Times* (London). (28 June 1845): 8.

Bryden, D. J. "Quality Control in the Making of Scientific Instruments: Kew Observatory and the Verification of Meteorological, Magnetic and Other Instruments, 1851–1899." *Bulletin of the Scientific Instrument Society*, no. 88 (2006): 48–59.

Bugge, Thomas, Kurt Moller Pedersen, and Peter de Clercq, eds. *An Observer of Observatories: The Journal of Thomas Bugge's Tour of Germany, Holland and England in 1777*. Aarhus: Aarhus University Press, 2010.

Burley, J., and K. Plenderleith, eds. *A History of the Radcliffe Observatory Oxford: The Biography of a Building*. Oxford: Green College, 2005.

Burton, James. "The History of the British Meteorological Office to 1905." PhD diss., UK: Open University, 1988.

Burton, Jim. "Robert FitzRoy and the Early History of the Meteorological Office." *British Journal for the History of Science* 19 (1986): 147–76.

Burton, Jim. "Scott, Robert Henry (1833–1916)." In *Oxford Dictionary of National Biography*. Oxford: Oxford University Press, 2004.

Burton, Jim. "Shaw, Sir (William) Napier (1854–1945)." In *Oxford Dictionary of National Biography*. Oxford: Oxford University Press, 2004.

Buttmann, Gunther. *The Shadow of the Telescope, A Biography of John Herschel*. Guildford, UK: Lutterworth, 1974.

C. V. B., C[harles] [Chree], and H[erbert] H[all] T[urner]. "G. W. Walker, 1874–1921." *Proceedings of the Royal Society A* 102 (1922–1923): xxii–xxvi.

Cahan, David. "The Geopolitics and Architectural Design of a Metrological Laboratory: The Physikalisch-Technische Reichsanstalt in Imperial Germany." In *The Development of the Laboratory: Essays on the Place of Experiment in Industrial Civilization*, edited by Frank A. J. L. James, 137–54. Basingstoke: Macmillan Press, 1989.

Cahan, David. *An Institute for an Empire: The Physikalisch-Technische Reichsanstalt 1871–1918*. Cambridge: Cambridge University Press, 1989.

Cannon, Susan Faye. *Science in Culture: The Early Victorian Period*. New York: Dawson and Science History Publications, 1978.

Cantor, G. N., and M. J. S. Hodge. *Conceptions of Ether: Studies in the History of Ether Theories 1740–1900*. Cambridge: Cambridge University Press, 1981.

Cardwell, D. S. L. *The Organisation of Science in England*. London: Heinemann, 1972.

Carrington, R. C. "Description of a Singular Appearance Seen in the Sun on September 1, 1859." *Monthly Notices of the Royal Astronomical Society* 20 (1859): 13–15.

Carter, Christopher. "Magnetic fever: Global Imperialism and Empiricism in the Nineteenth Century." *Transactions of the American Philosophical Society* 99, part 4 (2009): i–xxvi and 1–168.

Cattermole, M. J., and A. F. Wolfe. *Horace Darwin's Shop: A History of the Cambridge Scientific Instrument Company*. Bristol: Adam Hilger, 1987.

Cawood, John. "The Magnetic Crusade: Science and Politics in Early Victorian Britain." *Isis* 70 (1979): 492–518.

Cawood, John. "Terrestrial Magnetism and the Development of International Collaboration in the Early Nineteenth Century." *Annals of Science* 34 (1977): 551–87.

Chang, Hasok. *Inventing Temperature: Measurement and Scientific Progress*. Oxford: Oxford University Press, 2004.

Chapman, Allan. "Private Research and Public Duty: George Biddell Airy and the Search for Neptune." *Journal for the History of Astronomy* xix (1988): 121–39.

Chapman, Allan. "Science and the Public Good: George Biddell Airy (1801–92) and the Concept of a Scientific Civil Servant." In *Science, Politics and the Public Good*, edited by Nicolaas A. Rupke, 36–62. Basingstoke: Macmillan, 1988.

Chapman, Allan. "Sir George Airy (1801–1892) and the Concept of International Standards in Science, Timekeeping and Navigation." *Vistas in Astronomy* xxviii (1985): 321–28.

Chapman, Allan. *The Victorian Amateur Astronomer: Independent Astronomical Research in Britain 1820–1920*. Chichester: Wiley/Praxis, 1998.

Charbonneau, Paul. "The Rise and Fall of the First Solar Cycle Model." *Journal for the History of Astronomy* xxxiii (2002): 351–72.

Chree, Charles. "Description of the Kew Observatory." In *The Record of the Royal Society of London*, 137–53. London: Harrison & Sons, 1897.

Chree, Charles. "Observations on Atmospheric Electricity at the Kew Observatory." *Proceedings of the Royal Society* 60 (1896): 96–132.

Clark, Stuart. *The Sun Kings*. Princeton: Princeton University Press, 2007.

Clerke, A. M., revised by Joseph Gross. "Breen, James (1826–1866)." In *Oxford Dictionary of National Biography*. Oxford: Oxford University Press, 2004.

Clerke, A. M., revised by Anita McConnell. "Whipple, George Mathews (1842–1893)." In *Oxford Dictionary of National Biography*. Oxford: Oxford University Press, 2004.

Cotter, Charles H. "George Biddell Airy and His Mechanical Correction of the Magnetic Compass." *Annals of Science* 33 (1976): 263–74.

Crease, Robert P. *World in the Balance: The Historic Quest for an Absolute System of Measurement*. New York: W Norton & Company, 2011.

Crowe, Michael J., ed. *A Calendar of the Correspondence of Sir John Herschel*. With the assistance of David R. Dyck and James R. Kevin. Cambridge: Cambridge University Press, 1998.

"Cruel Cuts at the Meteorological Office." *Journal of Meteorology* 5 (1980): 137.

Davis, John L. "Weather Forecasting and the Development of Meteorological Theory at the Paris Observatory, 1853–1878." *Annals of Science* 41 (1984): 359–82.

De La Rue, Warren. "The Bakerian Lecture: On the Total Solar Eclipse of July 18th, 1860, Observed at Rivabellosa, Near Miranda de Ebro, in Spain." *Philosophical Transactions of the Royal Society* 152 (1862): 333–416.

De La Rue, Warren, [Balfour] Stewart and [Benjamin] Loewy. "A Comparison of the Kew Results of Observations on Sun-Spots with Those of Hofrath Schwabe, in Dessau, for the Year 1865." *Monthly Notices of the Royal Astronomical Society* 26 (1866): 76–77.

De La Rue, Warren. "Report on the Present State of Celestial Photography in England." In *Report of the Twenty-Ninth Meeting of the British Association for the Advancement of Science; Held at Aberdeen in September 1859*, 130–53. London: John Murray, 1860.

De La Rue, Warren. "Report on the Progress of Celestial Photography since the Aberdeen Meeting." In *Report of the Thirty-First Meeting of the British Association for the Advancement of Science, Held at Manchester in September 1861*, 94–96. London: John Murray, 1862.

De La Rue, Warren. "Warren De La Rue, Esq., DCL, FRS, examined." In *Royal Commission on Scientific Instruction and the Advancement of Science. Eighth Report*, 300–306. 1875.

Dick, Steven J. "National Observatories: An Overview," *Journal for the History of Astronomy* xxii (1991): 1–3.

Dick, Steven J. *Sky and Ocean Joined: The U.S. Naval Observatory 1830–2000*. Cambridge: Cambridge University Press, 2003.

"Ebor." "Watch Rate Papers." *The Horological Journal* 23 (1880): 56.

Edgerton, David. *Warfare State*. Cambridge: Cambridge University Press, 2006.

Enebakk, Vidar. "Hansteen's Magnetometer and the Origin of the Magnetic Crusade." *British Journal for the History of Science* 47 (2014): 587–608.

Ensor, R. C. K. *England 1870–1914*. Oxford: Oxford University Press, 1936.

Evans, David S., Terence J. Deeming, Betty Hall Evans, and Stephen Goldfarb, eds. *Herschel at the Cape: Diaries and Correspondence of Sir John Herschel, 1834–1838.* (Austin: University of Texas Press, 1969.

Fanning, A. E. *Steady as She Goes: A History of the Compass Department of the Admiralty*. London: Her Majesty's Stationery Office, 1996.

Fara, Patricia. *Pandora's Breeches: Women, Science and Power in the Enlightenment*. London: Random House, 2004.

"Fifteenth Meeting of the British Association for the Advancement of Science." *The Athenaeum*, no. 922 (28 June 1845): 639–45.

Fleming, James Rodger. *Meteorology in America, 1800–1870*. Baltimore: Johns Hopkins University Press, 1990.

Forbes, James D. "Report upon the Recent Progress and Present State of Meteorology." In *Report of the First and Second Meetings of the British Association for the Advancement of Science; at York in 1831, and at Oxford in 1832: Including its Proceedings, Recommendations, and Transactions*, 196–258. London: John Murray, 1833.

Forbes, James D. "Supplementary Report on Meteorology." In *Report of the Tenth Meeting of the British Association for the Advancement of Science; Held at Glasgow in August 1840*, 37–156. London: John Murray, 1841.

Galton, Douglas. "Address by Sir Douglas Galton, KCB, DCL, FRS, President." In *Report of the Sixty-Fifth Meeting of the British Association for the Advancement of Science; Held at Ipswich in September 1895*, 29–35. London: John Murray, 1896.

Galton, Douglas. "On the Reichsanstalt, Charlottenburg, Berlin." In *Report of the Sixty-Fifth Meeting of the British Association for the Advancement of Science; Held at Ipswich in September 1895*, 606–8. London: John Murray, 1896.

Galton, Francis. "Description of the Process of Verifying Thermometers at the Kew Observatory." *Proceedings of the Royal Society* 26 (1877): 84–89.

Galton, Francis. *Memories of My Life*. London: Methuen, 1909.

Galvin, J. F. P. "Kew Observatory." *Weather* 58 (2003): 478–84.

Gassiot, John P. *Remarks on the Resignation of Sir Edward Sabine K.C.B. of the Presidency of the Royal Society* (pamphlet, privately printed, 1870).

Gassiot, John P. "Report of the Kew Committee of the British Association for the Advancement of Science for 1860–1861." In *Report of the Thirty-First Meeting of the British Association for the Advancement of Science, Held at Manchester in September 1861*, xxxiii–xxxviii. London: John Murray, 1862.

Geikie, Archibald. "Sir Alfred Bray Kempe, 1849–1922." *Proceedings of the Royal Society* 102 (1923): i–x.

Glaisher, James. *Travels in the Air*. London: Richard Bentley, 1871.

Glazebrook, R. T., revised by Anita McConnell. "Darwin, Horace." In *Oxford Dictionary of National Biography*. Oxford: Oxford University Press, 2004.

Glazebrook, Richard. *Early Days at The National Physical Laboratory. A Lecture Delivered at the Laboratory on March 23rd, 1933*. Teddington: National Physical Laboratory, 1933.

Good, Gregory A. "A Shift of View: Meteorology in John Herschel's Terrestrial Physics." In *Intimate Universality: Local and Global Themes in the History of Weather and Climate*, edited by James Rodger Fleming, Vladimir Jankovic, and Deborah R. Coen, 35–67. Sagamore Beach: Science History Publications/USA, 2006.

Good, Gregory A. "Sir Edward Sabine." In *Oxford Dictionary of National Biography*. Oxford: Oxford University Press, 2004.

Gooday, Graeme. "Lies, Damned Lies and Declinism: Lyon Playfair, the Paris 1867 Exhibition and the Contested Rhetorics of Scientific Education and Industrial Performance." In *The Golden Age: Essays in British Social and Economic History, 1850–1870*, edited by Ian Inkster, Colin Griffin, Jeff Hill, and Judith Rowbotham, 105–20. Aldershot: Ashgate, 2000.

Gooday, Graeme. "Precision Measurement and the Genesis of Physics Teaching Laboratories in Victorian Britain." PhD diss., University of Kent at Canterbury, 1989.

Gooday, Graeme. "Precision Measurement and the Genesis of Physics Teaching Laboratories in Victorian Britain." *British Journal for the History of Science* 23 (1990): 25–51.

Gooday, Graeme. "Sunspots, Weather and the Unseen Universe: Balfour Stewart's Anti-Materialist Representations of 'Energy' in British Periodicals." In *Science Serialized: Representations of the Sciences in Nineteenth-century Periodicals*, edited by G. Cantor and S. Shuttleworth, 111–47. Cambridge, MA: MIT Press, 2004.

Gooday, Graeme J. N., and Morris F. Low. "Technology Transfer and Cultural Exchange: Western Scientists and Engineers Encounter Late Tokugawa and Meiji Japan." *Osiris* 13 (1998): 99–128.

Grier, David Alan. *When Computers Were Human*. Princeton, NJ: Princeton University Press, 2005.

H. J. "Herbert Tomlinson—1845–1931." *Obituary Notices of Fellows of the Royal Society* 1 (1933): 89–91.

Hall, Marie Boas, *All Scientists Now: The Royal Society in the Nineteenth Century*. Cambridge: Cambridge University Press, 1984.

Harrison, D. N. "The British Radiosonde: Its Debt to Kew." *Meteorological Magazine* 98 (1969): 186–90.

Hartog, P. J., revised by A. J. Meadows. "Rue, Warren de la (1815–1889)." In *Oxford Dictionary of National Biography*. Oxford: Oxford University Press, 2004.

Hartog, P. J., revised by Graeme J. N. Gooday. "Stewart, Balfour (1828–1887)." In *Oxford Dictionary of National Biography*. Oxford: Oxford University Press, 2004.

Hartog, P. J., revised by Anita McConnell. "Welsh, John (1824–1859)." In *Oxford Dictionary of National Biography*. Oxford University Press, 2004.

Headlam, Maurice, and Ivor Thomas, revised by Alan Booth. "Heath, Sir Thomas Little (1861–1940)." In *Oxford Dictionary of National Biography*. Oxford: Oxford University Press, 2004.

Herbert-Gustar, A. L., and P. A. Nott. *John Milne: Father of Modern Seismology*. Tenterden, Kent: Paul Norbury Publications, 1980.

Herrmann, Dieter B. "An Exponential Law for the Establishment of Observatories in the Nineteenth Century." *Journal for the History of Astronomy* 4 (1973): 57–58.

Herschel, John Frederick William. *A Preliminary Discourse on the Study of Natural Philosophy*. London: Longman, Rees, Orme, Brown & Green, 1830. Reprint, Chicago: University of Chicago Press, 1987.

Herschel, Sir J. F. W. "Extract of Letter Respecting Mr. Griesbach's Communication on the Solar Spots." *Monthly Notices of the Royal Astronomical Society* 8 (1847): 14–15.

Herschel, Sir J. F. W. *Results of Astronomical Observations Made during the Years 1834, 5, 6, 7, 8, at the Cape of Good Hope; Being the Completion of a Telescopic Survey of the Whole Surface of the Visible Heavens, Commenced in 1825*. London: Smith, Elder & Co., 1847.

Herschel, Sir John F. W. "On the Application of Photography to Astronomical Observations (Letter from Sir John F. W. Herschel to Colonel Sabine)." *Monthly Notices of the Royal Astronomical Society* 15 (1855): 158–59.

Higgitt, Rebekah. "A British National Observatory: The Building of the New Physical Observatory at Greenwich, 1889–1898." *British Journal for the History of Science* 47 (2014):609–35.

Hodgson, R. "On a Curious Appearance Seen in the Sun." *Monthly Notices of the Royal Astronomical Society* 20 (1859): 15–16.

Hollis, H. P. "The Decade 1870–1880." In *History of the Royal Astronomical Society 1820–1920*, edited by J. L. E. Dreyer and H. H. Turner, 167–211. London: Royal Astronomical Society, 1923.

Hoskin, Michael. *Discoverers of the Universe: William and Caroline Herschel*. Princeton: Princeton University Press, 2011.

Howarth, O. J. R. *The British Association for the Advancement of Science: A Retrospect 1831–1931*. London: British Association for the Advancement of Science, 1931.

Howse, D. *Greenwich Observatory: The Royal Observatory at Greenwich and Herstmonceux 1675–1975. Vol. 3: The Buildings and Instruments*. London: Taylor & Francis, 1975.

Hufbauer, K. *Exploring the Sun: Solar Science since Galileo*. Baltimore: Johns Hopkins University Press, 1991.

Hughes, Jeff. "'Divine Right' or Democracy? The Royal Society 'Revolt' of 1935." *Notes and Records of the Royal Society* 64 (2010): S101–S117.

Hughes, Jeff. "Redefining the Context: Oxford and the Wider World of British Physics, 1900–1940." In *Physics in Oxford 1839–1939*, edited by Robert Fox and Graeme Gooday, 267–300. Oxford: Oxford University Press, 2005.

Hull, Andrew. "War of Words: The Public Science of the British Scientific Community and the Origins of the Department of Scientific and Industrial Research, 1914–16." *British Journal for the History of Science* 32 (1999): 461–81.

Hutchins, Roger. "Birt, William Radcliff (1804–1881)." In *Oxford Dictionary of National Biography*. Oxford: Oxford University Press, 2004.

Hutchins, Roger. *British University Observatories 1772–1939*. Aldershot: Ashgate, 2008.

Hutchins, Roger. "The King's Observatory, Richmond Park ('The Kew Observatory')" (unpublished paper, communicated privately by the author to Lee Macdonald, October 2012).

Jacobs, L. "The 200-Years' Story of Kew Observatory." *Meteorological Magazine* 98 (1969): 162–71.

Jankovic, Vladimir. "The End of Classical Meteorology, c. 1800." In *The History of Meteoritics and Key Meteorite Collections: Fireballs, Falls and Finds*, edited by G. J. H. McCall, A. J. Bowden, and R. J. Howarth, 91–99. Geological Society, London, Special Publications 256, 2006.

Jankovic, Vladimir. "Ideological Crests versus Empirical Troughs: John Herschel's and William Radcliffe Birt's Research on Atmospheric Waves, 1843–50." *British Journal for the History of Science* 31 (1998): 21–40.

Jankovic, Vladimir. *Reading the Skies: A Cultural History of English Weather, 1650–1820.* Manchester: Manchester University Press, 2000.

"Kew and Eskdale Muir Observatories and the Meteorological Office." *Symons's Meteorological Magazine* 45 (1910): 101–2.

Kim, Dong-Won. "J. J. Thomson and the Emergence of the Cavendish School, 1885–1990." *British Journal for the History of Science* 28 (1995): 191–226.

Kinder, Anthony J. "Edward Walter Maunder FRAS (1851–1928): His Life and Times." *Journal of the British Astronomical Association* 118 (2008): 21–42.

King, Henry C. *The History of the Telescope.* High Wycombe, UK: Charles Griffin & Company, 1955. Reprint, Mineola, NY: Dover Publications, 2003.

K[nobel], E. B. "Warren De La Rue." *Monthly Notices of the Royal Astronomical Society* 50 (1890): 155–64.

Lang, H. R., revised by John K. Bradley. "Whipple, Robert Stewart (1871–1953)." In *Oxford Dictionary of National Biography.* Oxford: Oxford University Press, 2004.

Lankford, John. "Amateurs and Astrophysics: A Neglected Aspect in the Development of a Scientific Specialty." *Social Studies of Science* 11 (1981): 275–303.

Lawrence, Christopher. "Incommunicable Knowledge: Science, Technology and the Clinical Art in Britain 1850–1914." *Journal of Contemporary History* 20 (1985): 503–20.

Le Conte, David. "Warren De La Rue—Pioneer Astronomical Photographer." *The Antiquarian Astronomer*, no. 5 (2011): 14–35.

Lightman, Bernard. "Huxley and the Devonshire Commission." In *Victorian Scientific Naturalism: Community, Identity, Continuity*, edited by Bernard Lightman and Gowan Dawson, 102–26. Chicago: University of Chicago Press, 2014.

Lindsay, E. M. "The Astronomical Instruments of HM King George III Presented to Armagh Observatory." *Irish Astronomical Journal* 9 (1969): 57–68.

Lodge, Oliver J. "Section A.—Mathematical and Physical Science. President of the Section—Professor Oliver J. Lodge, DSc, LLD, FRS [Address by Oliver Lodge, 20 August 1891]." In *Report of the Sixty-First Meeting of the British Association for the Advancement of Science; Held at Cardiff in August 1891*, 547–57. London: John Murray, 1892.

Longair, Malcolm. "Charles Thomson Rees Wilson." In *Oxford Dictionary of National Biography.* Oxford: Oxford University Press, 2004.

Macdonald, Lee T. "Making Kew Observatory: The Royal Society, the British Association and the Politics of Early Victorian Science." *British Journal for the History of Science* 48 (2015): 409–33.

Macdonald, Lee T. "'Solar Spot Mania': The Origins and Early Years of Solar Research at Kew Observatory, 1852–1860." *Journal for the History of Astronomy* 46 (2015): 469–90.

MacLeod, R. M. "Resources of Science in Victorian England: The Endowment of Science Movement, 1868–1900." In *Science and Society, 1600–1900*, edited by P. Mathias, 111–66. Cambridge: Cambridge University Press 1972.

MacLeod, Roy. "Introduction." In *Public Science and Public Policy in Victorian England*, vii–xiv. Aldershot, UK: Variorum, 1996.

MacLeod, Roy. "The Royal Society and the Government Grant: Notes on the Administration of Scientific Research, 1849–1914." *The Historical Journal* 14, no. 2 (1971): 323–58.

MacLeod, Roy M. "Science and the Treasury: Principles, Personalities and Policies, 1870–85." In *The Patronage of Science in the Nineteenth Century*, edited by G. L'E. Turner, 115–72. Leyden, Noordhoff International Publishing, 1976.

MacLeod, Roy. "The Support of Victorian Science: The Endowment of Research Movement in Great Britain, 1868–1900." *Minerva* 4 (1971): 197–230.

MacLeod, Roy. "Whigs and Savants: Reflections on the Reform Movement in the Royal Society, 1830–48." In *Metropolis and Province: Science in British Culture 1780–1850*, edited by Ian Inkster and Jack Morrell, 55–90. Philadelphia: University of Pennsylvania Press, 1983.

MacLeod, Roy. "The X-Club. A Social Network of Science in Late-Victorian England." *Notes and Records of the Royal Society* 24 (1969): 305–22.

MacLeod, Roy, and Peter Collins, eds. *The Parliament of Science: The British Association for the Advancement of Science 1831–1981*. Middlesex: Science Reviews, 1981.

Magnello, Eileen. *A Century of Measurement: An Illustrated History of the National Physical Laboratory*. Bath: Canopus Publishing, 2000.

Marder, Arthur J. *The Anatomy of British Sea Power: A History of British Naval Policy in the pre-Dreadnought Era, 1880–1905*. London: Frank Cass & Co., 1964.

Martin, D. C. "The Gassiot Committee of the Royal Society and Meteorological Research." *Nature* 190 (1961): 212–13.

Mason, B. J. "Simpson, Sir George Clarke (1878–1965)." In *Oxford Dictionary of National Biography*. Oxford: Oxford University Press, 2004.

[Mason, Basil]. "Foreword by the Director-General of the Meteorological Office." *Meteorological Magazine* 98 (1969): 161.

Massey, Sir Harrie, and M. O. Robins. *History of British Space Science*. Cambridge: Cambridge University Press, 1986.

Maunder, E. Walter. *The Royal Observatory Greenwich: A Glance at its History and Work*. London: The Religious Tract Society, 1900.

Mayes, Julian. "Kew Observatory, 1769 to 1980: Climatological Implications of an Observatory Closure," in *Observatories and Climatological Research: Proceedings of a Conference Held at St Mary's College, University of Durham, 5–7 September, 1991, in Celebration of 150 Years of Meteorology in Durham*, edited by Brian D. Giles and Joan M. Kenworthy, 152–162. Dept. of Geography, Durham University, UK, in collaboration with the Royal Meteorological Society, 1994.

McConnell, Anita. *King of the Clinicals: The Life and Times of J. J. Hicks (1837–1916)*. York: William Sessions, 1998.

Meadows A. J. *Greenwich Observatory: The Royal Observatory at Greenwich and Herstmonceux. Vol. 2: Recent History (1836–1975)*. London: Taylor and Francis, 1975.

Meadows A. J. *Science and Controversy: A Biography of Sir Norman Lockyer*, 2nd ed. Basingstoke: Macmillan, 2008.

Meadows, A. J., and J. E. Kennedy. "The Origin of Solar-Terrestrial Studies." *Vistas in Astronomy* 25 (1982): 419–26.

"The Meteorological Committee." *The Saturday Review* 26 (7 November 1868): 622–23.

"The Meteorological Department." *The Athenaeum*, no. 2136 (3 October 1868): 436–37.

Meteorological Office. *The Observatories' Year Book 1922.* London: Her Majesty's Stationery Office, 1925.

Meteorological Office. *The Observatories' Year Book 1965.* London: Her Majesty's Stationery Office, 1968.

Meteorological Office. *The Observatories' Year Book 1967.* London: Her Majesty's Stationery Office, 1969.

Middleton, W. E. Knowles. *The History of the Barometer.* Baltimore, MD: The Johns Hopkins Press, 1964.

Middleton, W. E. Knowles. *A History of the Thermometer and Its Use in Meteorology.* Baltimore, MD: The Johns Hopkins Press, 1966.

Miller, David Philip. "The Revival of the Physical Sciences in Britain, 1815–1840." *Osiris,* 2nd series, no. 2 (1986):107–34.

Minutes of Evidence Taken before the Committee Appointed by the Treasury to Consider the Desirability of Establishing a National Physical Laboratory; with Appendices and Index. London: Her Majesty's Stationery Office, 1898.

Morrell, Jack, and Arnold Thackray, eds. *Gentlemen of Science: Early Correspondence of the British Association for the Advancement of Science.* London: Royal Historical Society, 1984.

Morrell, Jack, and Arnold Thackray. *Gentlemen of Science: Early Years of the British Association for the Advancement of Science.* Oxford: Clarendon Press, 1981.

Morrison-Low, A. D. "Women in the Nineteenth-Century Scientific Instrument Trade." In *Science and Sensibility: Gender and Scientific Enquiry, 1780–1945,* edited by Marina Benjamin, 89–117. Oxford: Blackwell, 1991.

Morton, Alan Q., and Wess, Jane A. *Public & Private Science: The King George III Collection.* Oxford: Oxford University Press, in association with the Science Museum, 1993.

Morus, Iwan Rhys. *When Physics Became King.* Chicago: University of Chicago Press, 2005.

Mörzer-Bruyns, W. F. J. *Sextants at Greenwich: A Catalogue of the Mariner's Quadrants, Mariner's Astrolabes, Cross-staffs, Backstaffs, Octants, Sextants, Quintants, Reflecting Circles, and Artificial Horizons in the National Maritime Museum, Greenwich.* Oxford and London: Oxford University Press and National Maritime Museum, 2009.

Moseley, Russell. "Glazebrook, Sir Richard Tetley (1854–1935)." In *Oxford Dictionary of National Biography.* Oxford: Oxford University Press, 2004.

Moseley, Russell. "The Origins and Early Years of the National Physical Laboratory: A Chapter in the Pre-history of British Science Policy." *Minerva* 16 (1978): 222–50.

Moseley, Russell. "Science, Government & Industrial Research: The Origins & Development of the National Physical Laboratory, 1900–1975." PhD diss., University of Sussex, 1976.

"Mr. John Welsh." In "Obituary Notices of Deceased Fellows." *Proceedings of the Royal Society*, no. 10 (1859–1860): xxxiv–xxxviii.

"Mr. T. H. Blakesley." *Nature* 123 (2 March 1929): 324.

Musselman, Elizabeth Green. "Swords into Ploughshares: John Herschel's Progressive View of Astronomical and Imperial Governance." *British Journal for the History of Science* 31 (1998): 419–36.

"The National Physical Laboratory. Report on the Observatory Department for the Year Ending December 31, 1900." *Proceedings of the Royal Society* 68 (1901): 421–53.

The National Physical Laboratory. Report for the Year 1901. London: National Physical Laboratory, 1902.

The National Physical Laboratory. Report for the Year 1902. London: National Physical Laboratory, 1903.

The National Physical Laboratory. Report for the Year 1903. London: National Physical Laboratory, 1904.

The National Physical Laboratory. Report for the Year 1905. Teddington: National Physical Laboratory, 1906.

The National Physical Laboratory. Report for the Year 1906. Teddington: National Physical Laboratory, 1907.

The National Physical Laboratory. Report for the Year 1907. Teddington: National Physical Laboratory, 1908.

The National Physical Laboratory. Report for the Year 1908. Teddington: National Physical Laboratory, 1909.

The National Physical Laboratory. Report for the Year 1909. Teddington: National Physical Laboratory, 1910.

The National Physical Laboratory. Report for the Year 1912. Teddington: National Physical Laboratory, 1913.

The National Physical Laboratory. Report for the Year 1913–14. Teddington: National Physical Laboratory, 1914.

Olesko, Kathryn M. "Physics and Metrology." In *The Oxford Handbook of the History of Physics*, edited by Jed Buchwald and Robert Fox, 698–718. Oxford: Oxford University Press, 2013.

"On the Establishment of a National Physical Laboratory." In *Report of the Sixty-Sixth Meeting of the British Association for the Advancement of Science Held at Liverpool in September 1896*, 82–87. London: John Murray, 1897.

Orange, A. D. "The Beginnings of the British Association, 1831–1851." In *The Parliament of Science: The British Association for the Advancement of Science*, edited by Roy MacLeod and Peter Collins, 43–64. Northwood: Science Reviews, 1981.

Osterbrock, Donald E., John R. Gustafson, and W. J. Shiloh Unruh. *Eye on the Sky: Lick Observatory's First Century.* Berkeley: University of California Press, 1988.

Pang, Alex Soojung-Kim. *Empire and the Sun: Victorian Solar Eclipse Expeditions.* Stanford: Stanford University Press, 2002.

Panwitz, Hrsg. von Sebastian, and Ingo Schwarz. *Alexander von Humboldt Familie Mendelssohn Briefwechsel.* Berlin: Akademie Verlag GmbH, 2011.

Pearson, Karl. *The Life, Letters and Labours of Francis Galton, Vol. 2: Researches of Middle Life*. Cambridge: Cambridge University Press, 1924.

Peden, G. C. "Murray, Sir George Herbert (1849–1936)." In *Oxford Dictionary of National Biography*. Oxford: Oxford University Press, 2004.

Peden, G. C. *The Treasury and British Public Policy, 1906–1959*. Oxford: Oxford University Press, 2000.

Pyatt, Edward. *The National Physical Laboratory: A History*. Bristol: Adam Hilger, 1983.

Quill, Humphrey. *John Harrison: The Man Who Found Longitude*. London: John Baker, 1966.

Ratcliff, Jessica. *The Transit of Venus Enterprise in Victorian Britain*. London: Pickering & Chatto, 2008.

Ratcliff, Jessica. "Travancore's Magnetic Crusade: Geomagnetism and the Geography of Scientific Production in a Princely State." *British Journal for the History of Science* 49 (2016): 325–52.

Read W. J. "History of the Firm Negretti & Zambra." *Bulletin of the Scientific Instrument Society*, no. 5 (1985): 8–10.

The Record of the Royal Society of London for the Promotion of Natural Knowledge, 4th ed. London: Royal Society, 1940.

Reid, William. *"We're Certainly Not Afraid of Zeiss": Barr & Stroud Binoculars and the Royal Navy*. Edinburgh: National Museums of Scotland Publishing, 2001.

Reidy, Michael S. *Tides of History: Ocean Science and Her Majesty's Navy*. Chicago: University of Chicago Press, 2008.

Reingold, Nathan. "Sabine, Edward." In Vol. 12, *Dictionary of Scientific Biography*, edited by Charles Coulston Gillispie, 49–53. New York: Charles Scribner's Sons, 1975.

Report of the Committee Appointed by the Treasury to Consider the Desirability of Establishing a National Physical Laboratory. London: Her Majesty's Stationery Office, 1898.

"Report of the Council. Presented to the Royal Society, November 30, 1911." In *Year-Book of the Royal Society 1912*, 182–89. London: Royal Society, 1912.

Report of the Fifteenth Meeting of the British Association for the Advancement of Science, Held at Cambridge, 1845. London: John Murray, 1846.

Report of the Fortieth Meeting of the British Association for the Advancement of Science; Held at Liverpool in September 1870. London: John Murray, 1871.

"Report of the Incorporated Kew Committee for the Year Ending December 31, 1893." *Proceedings of the Royal Society* 55 (1893): 307–39.

"Report of the Incorporated Kew Committee for the Year Ending December 31, 1894." *Proceedings of the Royal Society* 57 (1894): 500–529.

"Report of the Kew Committee for the Fifteen Months Ending October 31, 1872." *Proceedings of the Royal Society* 21 (1872): 40–46.

"Report of the Kew Committee for the Fourteen Months Ending December 31, 1891." *Proceedings of the Royal Society* 51 (1891): 152–82.

"Report of the Kew Committee for the Year Ending December 31, 1892." *Proceedings of the Royal Society* 53 (1892): 322–51.

"Report of the Kew Committee for the Year Ending October 31, 1875." *Proceedings of the Royal Society* 24 (1875): 102–15.

"Report of the Kew Committee for the Year Ending October 31, 1876," *Proceedings of the Royal Society* 25 (1876): 370–84.

"Report of the Kew Committee for the Year Ending October 31, 1879." *Proceedings of the Royal Society* 29 (1879): 445–64.

"Report of the Kew Committee for the Year Ending October 31, 1880." *Proceedings of the Royal Society* 31 (1880): 115–36.

"Report of the Kew Committee for the Year Ending October 31, 1881." *Proceedings of the Royal Society* 33 (1881): 80–99.

"Report of the Kew Committee for the Year Ending October 31, 1882." *Proceedings of the Royal Society* 34 (1882): 343–65.

"Report of the Kew Committee for the Year Ending October 31, 1884." *Proceedings of the Royal Society* 37 (1884): 462–83.

"Report of the Kew Committee for the Year Ending October 31, 1885." *Proceedings of the Royal Society* 39 (1885): 314–38.

"Report of the Kew Committee for the Year Ending October 31, 1886." *Proceedings of the Royal Society* 41 (1886): 400–422.

"Report of the Kew Committee for the Year Ending October 31, 1887." *Proceedings of the Royal Society* 43 (1887): 211–34.

"Report of the Kew Committee for the Year Ending October 31, 1888." *Proceedings of the Royal Society* 45 (1888): 73–98.

"Report of the Kew Committee for the Year Ending October 31, 1889." *Proceedings of the Royal Society* 46 (1889): 474–95.

"Report of the Kew Committee for the Year Ending October 31, 1890." *Proceedings of the Royal Society* 48 (1890): 491–504.

"Report of the Kew Observatory Committee for the Year ending December 31, 1895." *Proceedings of the Royal Society* 59 (1895): 383–414.

"Report of the Kew Observatory Committee for the Year Ending December 31, 1896." *Proceedings of the Royal Society* 61 (1896): 96–126.

"Report of the Kew Observatory Committee for the Year Ending December 31, 1897." *Proceedings of the Royal Society* 63 (1897): 161–91.

"Report of the Kew Observatory Committee for the Year Ending December 31, 1898." *Proceedings of the Royal Society* 65 (1898): 1–36.

"Report of the Kew Observatory Committee for the Year Ending December 31, 1899." *Proceedings of the Royal Society* 66 (1899): 341–73.

Report of the Sixteenth Meeting of the British Association for the Advancement of Science, Held at Southampton in September 1846. London: John Murray, 1847.

Report of the Sixty-Second Meeting of the British Association for the Advancement of Science; Held at Edinburgh in August 1892. London: John Murray, 1893.

Report of the Sixty-Sixth Meeting of the British Association for the Advancement of Science; Held at Liverpool in September 1896. London: John Murray, 1897.

Report of the Thirtieth Meeting of the British Association for the Advancement of Science; Held at Oxford in June and July 1860. London: John Murray, 1861.

Report of the Thirty-Eighth Meeting of the British Association for the Advancement of Science; Held at Norwich in August 1868. London: John Murray, 1869.

Report of the Thirty-Ninth Meeting of the British Association for the Advancement of Science; Held at Exeter in August 1869. London: John Murray, 1870.

Report of the Thirty-Second Meeting of the British Association for the Advancement of Science; Held at Cambridge in October 1862. London: John Murray, 1863.

Report of the Thirty-Third Meeting of the British Association for the Advancement of Science; Held at Newcastle-Upon-Tyne in August and September 1863. London: John Murray, 1864.

Report of the Twelfth Meeting of the British Association for the Advancement of Science, Held at Manchester, 1842. London: John Murray, 1843.

Report of the Twenty-Eighth Meeting of the British Association for the Advancement of Science; Held at Leeds in September 1858. London: John Murray, 1859.

Report of the Twenty-Fourth Meeting of the British Association for the Advancement of Science, Held at Liverpool in September 1854. London: John Murray, 1855.

Report of the Twenty-Ninth Meeting of the British Association for the Advancement of Science; Held at Aberdeen in September 1859. London: John Murray, 1860.

Rigaud, Gibbes. "Dr Demainbray and the King's Observatory at Kew." *The Observatory* 5 (1882): 279–85.

Rolt, L. T. C. *The Aeronauts: A History of Ballooning, 1783–1903.* London: Longman, 1966.

Ronalds, Beverley F. *Sir Francis Ronalds: Father of the Electric Telegraph.* London: Imperial College Press, 2016.

Ronalds, Francis. "On Photographic Self-Registering Meteorological and Magnetical Instruments." *Philosophical Transactions of the Royal Society* 137 (1847): 111–17.

Ronalds, Francis. "Report concerning the Observatory of the British Association at Kew, from August the 1st, 1843, to July the 31st, 1844." In *Report of the Fourteenth Meeting of the British Association for the Advancement of Science, Held at York, 1844.* London: John Murray, 1845: 120–31.

Ross, Sydney. "*Scientist*: The Story of a Word." *Annals of Science* 18 (1962): 65–85.

Rothermel, Holly. "Images of the Sun: Warren De La Rue, George Biddell Airy and Celestial Photography." *British Journal for the History of Science* 26 (1993): 137–69.

"Royal Astronomical Society. Session 1871–72. Sixth Meeting, April 12th, 1872." *The Astronomical Register* 10 (1872): 111–21.

Royal Commission on Scientific Instruction and the Advancement of Science. Eighth Report, 1875.

Ruskin, Steven. *John Herschel's Cape Voyage: Private Science, Public Imagination and the Ambitions of Empire.* Aldershot, UK: Ashgate, 2004.

Savours, Ann, and Anita McConnell. "The History of the Rossbank Observatory, Tasmania." *Annals of Science* 39 (1982): 527–64.

Schaaf, Larry J. *Out of the Shadows: Herschel, Talbot and the Invention of Photography.* New Haven: Yale University Press, 1992.

Schaffer, Simon. "Astronomers Mark Time: Discipline and the Personal Equation." *Science in Context*, no. 2 (1988): 115–45.

Schaffer, Simon. "Keeping the Books at Paramatta Observatory." In *The Heavens on Earth: Observatories and Astronomy in Nineteenth-Century Science and Culture*, edited by David Aubin, Charlotte Bigg, and H. Otto Sibum, 118–147. Durham, NC: Duke University Press, 2010.

Schaffer, Simon. "Late Victorian Metrology and Its Instrumentation: A Manufactory of Ohms." In *Invisible Connections: Instruments, Institutions, and Science*, edited by Robert Bud and Susan E. Cozzens, 23–56. Bellingham, WA: SPIE Optical Engineering Press, 1992.

Schaffer, Simon. "Metrology, Metrication and Victorian Values." In *Victorian Science in Context*, edited by Bernard Lightman, 438–74. Chicago: University of Chicago Press, 1997.

Schaffer, Simon. "Where Experiments End: Tabletop Trials in Victorian Astronomy." In *Scientific Practice: Theories and Stories of Doing Physics*, edited by Jed Z. Buchwald, 257–99. Chicago: University of Chicago Press, 1995.

Schuster, A. "Memoir of the Late Professor Balfour Stewart, LL.D., F.R.S." In *Memoirs and Proceedings of the Manchester Literary & Philosophical Society*, 4th series, no. 1 (1888): 253–72.

Schuster, Arthur. *Biographical Fragments*. London: Macmillan, 1932.

"Scientific Balloon Ascent." *Illustrated London News*, no. 576 (28 August 1852): 175.

"Scientific Balloon Ascent." *Illustrated London News*, no. 577 (4 September 1852): 192.

Scott, Robert Henry. "The History of the Kew Observatory." *Proceedings of the Royal Society* 39 (1885): 37–86.

Scrase, F. J. "Some Reminiscences of Kew Observatory in the Twenties." *Meteorological Magazine* 98 (1969): 180–86.

Shaw, Sir Napier. "An Episode in the History of Kew Observatory." *The Meteorological Magazine* 61 (1926): 125–29.

Siegel, Daniel. "Balfour Stewart and Gustav Robert Kirchhoff: Two Independent Approaches to Kirchhoff's Radiation Laws." *Isis* 67 (1976): 565–600.

S[impson], G. C. "Charles Chree, 1860–1928." *Proceedings of the Royal Society*, A, 122 (1929): vii–xiv.

"Sir Edward Sabine." In "Obituary Notices of Fellows Deceased." *Proceedings of the Royal Society* 51 (1892): 308-314 i–li.

Smith, Crosbie. *The Science of Energy: A Cultural History of Energy Physics in Victorian Britain*. London: The Athlone Press, 1998.

Smith, Robert W. "The Cambridge Network in Action: The Discovery of Neptune." *Isis* 80 (1989): 395–422.

Smith, Robert W. "A National Observatory Transformed: Greenwich in the Nineteenth Century." *Journal for the History of Astronomy* xxii (1991): 5–20.

Smith, Robert W. "Remaking Astronomy: Instruments and Practice in the Nineteenth and Twentieth Centuries." In *The Cambridge History of Science. Vol. 5: The Modern Physical and Mathematical Sciences*, edited by Mary Jo Nye, 154–73. Cambridge: Cambridge University Press, 2003.

Stewart, K. H. "Development of Rocket and Satellite Experiments at Kew Observatory, 1959–61." *Meteorological Magazine* 98 (1969): 195–96.

Stewart, Balfour. "On the Great Magnetic Disturbance of August 28 to September 7, 1859, as Recorded by Photography at the Kew Observatory." *Proceedings of the Royal Society* 11 (1861): 407–10.

Stewart, Balfour. "On the Nature of Those Red Protuberances Which Are Seen on the Sun's Limb during a Total Eclipse." *Philosophical Magazine*, series 4, no. 24 (1862): 302–5.

Stewart, Balfour. "On Sun-Spots and Their Connection with Planetary Configurations." *Transactions of the Royal Society of Edinburgh* 23 (1864): 499–504.

Stewart, Balfour. "Physical Meteorology. I. Its Present Position." *Nature* 1 (1869): 101–3.

Stewart, Balfour, and J. Norman Lockyer. "The Sun as a Type of the Material Universe. Part II. The Place of Life in a Universe of Energy." *Macmillan's Magazine* 18 (1868): 319–27.

Stewart, Balfour, and Peter Guthrie Tait. "On the Heating of a Disc by Rapid Rotation in Vacuo." *Proceedings of the Royal Society* 14 (1865): 393–403.

Stewart, Balfour, and Peter Guthrie Tait. "Preliminary Note on the Radiation from a Revolving Disc." *Proceedings of the Royal Society* 14 (1865): 90.

Strange, Alexander. "Lieutenant-Colonel Alexander Strange, FRS, Examined." In *Royal Commission on Scientific Instruction and the Advancement of Science. Eighth Report* 75–92. 1875.

Strange, Alexander. "On the Insufficiency of Existing National Observatories." *Monthly Notices of the Royal Astronomical Society* 32 (1872): 238–41.

Strange, Lieut.-Col. A. "On the Necessity for State Intervention to Secure the Progress of Physical Science." In *Report of the Thirty-Eighth Meeting of the British Association for the Advancement of Science; Held at Norwich in August 1868*, 6–8. London: John Murray, 1869.

Sviedrys, Romualdas. "The Rise of Physical Science at Victorian Cambridge." *Historical Studies in the Physical Sciences* 2 (1970): 127–52.

Sviedrys, Romualdas. "The Rise of Physics Laboratories in Britain." *Historical Studies in the Physical Sciences* 7 (1976): 405–36.

Sweetman, John, and Anita McConnell. "Brisbane, Sir Thomas Makdougall, Baronet (1773–1860)." In *Oxford Dictionary of National Biography*. Oxford: Oxford University Press, 2004.

Symons, Eleanor Putnam. "Ronalds, Sir Francis (1788–1873)." In *Oxford Dictionary of National Biography*. Oxford: Oxford University Press, 2004.

T[ait], P[eter] G[uthrie]. "Dr. Balfour Stewart." In "Obituary Notices of Fellows Deceased." *Proceedings of the Royal Society* 46 (1889): ix–xi.

Taylor, Geoffrey I., and E. H. E. Havelock. "William Cecil Dampier 1867–1952." *Obituary Notices of Fellows of the Royal Society* 9 (1954): 55–63.

Thomson, William. "Address of Sir William Thomson, Knt., LL.D., F.R.S., President." In *Report of the Forty-First Meeting of the British Association for the Advancement of Science; Held at Edinburgh in August 1871*, lxxxiv–cv. London: John Murray, 1872.

Thomson, William. "Sir William Thomson, FRS, Further Examined." In *Royal Commission on Scientific Instruction and the Advancement of Science. Eighth Report*, 105–14. 1875.

Tobin, William. *The Life and Science of Léon Foucault: The Man Who Proved the Earth Rotates.* Cambridge: Cambridge University Press, 2003.

Tombs, Robert. *France 1814–1914.* Harlow: Longman, 1996.

A Treatise on Meteorological Instruments. London: Negretti & Zambra, 1864.

Tucker, Jennifer. "Photography as Witness, Detective, and Impostor: Visual Representation in Victorian Science." In *Victorian Science in Context*, edited by Bernard Lightman, 378–408. Chicago: University of Chicago Press, 1997.

Tucker, Jennifer. "Voyages of Discovery on Oceans of Air: Scientific Observation and the Image of Science in an Age of 'Balloonacy.'" *Osiris* 11 (1996): 144–76.

Van Allen, James A., and Frances Bagenal. "Planetary Magnetospheres and the Interplanetary Medium." In *The New Solar System*, 4th ed., edited by J. Kelly Beatty, Carolyn Collins Petersen, and Andrew Chaikin, 39–58. Cambridge, MA, and Cambridge, UK: Sky Publishing and Cambridge University Press, 1999.

Walker, Malcolm. *History of the Meteorological Office*. Cambridge: Cambridge University Press, 2012.

Wallace, Gordon. "Meteorological Observation at the Radcliffe Observatory." In *A History of the Radcliffe Observatory Oxford: The Biography of a Building*, edited by J. Burley and K. Plenderleith, 103–28. Oxford: Green College, 2005.

Warwick, Andrew. *Masters of Theory: Cambridge and the Rise of Mathematical Physics*. Chicago: University of Chicago Press, 2005.

Welsh, John. "An Account of Meteorological Observations in Four Balloon Ascents, Made under the Direction of the Kew Observatory Committee of the British Association for the Advancement of Science." *Philosophical Transactions of the Royal Society* 143 (1853): 311–46.

Whipple, F. J. W. "Some Aspects of the Early History of Kew Observatory." *Quarterly Journal of the Royal Meteorological Society* 63 (1937): 127–35.

White, Walter. *The Journals of Walter White*. London: Chapman & Hall, 1898.

"William Radcliff Birt." *The Astronomical Register* 20 (1882b): 12–13.

"William Radcliff Birt." *Monthly Notices of the Royal Astronomical Society* 42 (1882a): 142–44.

Williams, Mari E. W. *The Precision Makers: A History of the Instruments Industry in Britain and France, 1870–1939*. London: Routledge, 1994.

Wilson, David B. "Stokes and Kelvin, Cambridge and Glasgow, light and heat." In *Cambridge Scientific Minds*, edited by Peter Harman and Simon Mitton, 107–22. Cambridge: Cambridge University Press, 2002.

Wise, M. Norton. "Introduction." In *The Values of Precision*, edited by M. Norton Wise, 3–13. Princeton, NJ: Princeton University Press, 1995.

Woodward, Sir Llewellyn. *The Age of Reform 1815–1870*, 2nd ed. Oxford: Oxford University Press, 1962.

Year-Book of the Royal Society of London. 1912. London: Royal Society, 1912.

Year Book of the Royal Society 1969. London: Royal Society, 1969.

Year Book of the Royal Society 1981. London: Royal Society, 1981.

INDEX